Supersymmetry in Quantum Mechanics

Supersymmetry in Quantum Mechanics

Fred Cooper
Los Alamos National Laboratory

Avinash Khare
Institute of Physics, Bhubaneswar

Uday Sukhatme
University of Illinois, Chicago

World Scientific
Singapore • New Jersey • London • Hong Kong

Published by

World Scientific Publishing Co. Pte. Ltd.

P O Box 128, Farrer Road, Singapore 912805

USA office: Suite 1B, 1060 Main Street, River Edge, NJ 07661

UK office: 57 Shelton Street, Covent Garden, London WC2H 9HE

British Library Cataloguing-in-Publication Data
A catalogue record for this book is available from the British Library.

ISBN 981-02-4605-6
ISBN 981-02-4612-9 (pbk)

Printed in Singapore by Uto-Print

DEDICATED TO OUR WIVES
CATHERINE, PUSHPA AND MEDHA.

Preface

During the past fifteen years, a new conceptual framework for understanding potential problems in quantum mechanics has been developed using ideas borrowed from quantum field theory. The concept of supersymmetry when applied to quantum mechanics has led to a new way of relating Hamiltonians with similar spectra. These ideas are simple enough to be a part of the physics curriculum.

The aim of this book is to provide an elementary description of supersymmetric quantum mechanics which complements the traditional coverage found in existing quantum mechanics textbooks. In this spirit we give problems at the end of each chapter as well as complete solutions to all the problems. While planning this book, we realized that it was not possible to cover all the recent developments in this field. We therefore decided that, instead of pretending to be comprehensive, it was better to include those topics which we consider important and which could be easily appreciated by students in advanced undergraduate and beginning graduate quantum mechanics courses.

It is a pleasure to thank all of our many collaborators who helped in our understanding of supersymmetric quantum mechanics. This book could not have been written without the love and support of our wives Catherine, Pushpa and Medha.

Fred Cooper, Avinash Khare, Uday Sukhatme
Los Alamos, Bhubaneswar, Chicago
September 2000

Contents

Chapter 1

Introduction

Supersymmetry (SUSY) arose as a response to attempts by physicists to obtain a unified description of all basic interactions of nature. SUSY relates bosonic and fermionic degrees of freedom combining them into superfields which provides a more elegant description of nature. The algebra involved in SUSY is a graded Lie algebra which closes under a combination of commutation and anti-commutation relations. It may be noted here that so far here has been no experimental evidence of SUSY being realized in nature. Nevertheless, in the last fifteen years, the ideas of SUSY have stimulated new approaches to other branches of physics like atomic, molecular, nuclear, statistical and condensed matter physics as well as nonrelativistic quantum mechanics. Naively, unbroken SUSY leads to a degeneracy between the spectra of the fermions and bosons in a unified theory. Since this is not observed in nature one needs SUSY to be spontaneously broken. It was in he context of trying to understand the breakdown of SUSY in field theory hat the whole subject of SUSY quantum mechanics was first studied.

Once people started studying various aspects of supersymmetric quantum mechanics (SUSY QM), it was soon clear that this field was interesting in its own right, not just as a model for testing field theory methods. It was realized that SUSY QM gives insight into the factorization method of Infeld and Hull which was the first attempt to categorize the analytically solvable potential problems. Gradually a whole technology was evolved based on SUSY to understand the solvable potential problems and even to discover new solvable potential problems. One purpose of this book is to introduce and elaborate on the use of these new ideas in unifying how one looks at solving bound state and continuum quantum mechanics problems.

Let us briefly mention some consequences of supersymmetry in quantum mechanics. It gives us insight into why certain one-dimensional potentials are analytically solvable and also suggests how one can discover new solvable potentials. For potentials which are not exactly solvable, supersymmetry allows us to develop an array of powerful new approximation methods. In this book, we review the theoretical formulation of SUSY QM and discuss how SUSY helps us find exact and approximate solutions to many interesting quantum mechanics problems.

We will show that the reason certain potentials are exactly solvable can be understood in terms of a few basic ideas which include supersymmetric partner potentials and shape invariance. Familiar solvable potentials all have the property of shape invariance. We will also use ideas of SUSY to explore the deep connection between inverse scattering and isospectral potentials related by SUSY QM methods. Using these ideas we show how to construct multi-soliton solutions of the Korteweg-de Vries (KdV) equation. We then turn our attention to introducing approximation methods that work particularly well when modified to utilize concepts borrowed from SUSY. In particular we will show that a supersymmetry inspired WKB approximation is exact for a class of shape invariant potentials. Supersymmetry ideas also give particularly nice results for the tunneling rate in a double well potential and for improving large N expansions and variational methods.

In SUSY QM, one is considering a simple realization of a SUSY algebra involving bosonic and fermionic operators which obey commutation and anticommutation relations respectively. The Hamiltonian for SUSY QM is a 2×2 matrix Hamiltonian which when diagonalized gives rise to 2 separate Hamiltonians whose eigenvalues, eigenfunctions and S-matrices are related because of the existence of fermionic operators which commute with the Hamiltonian. These relationships will be exploited to categorize analytically solvable potential problems. Once the algebraic structure is understood, the results follow and one never needs to return to the origin of the Fermi-Bose symmetry. The interpretation of SUSY QM as a degenerate Wess-Zumino field theory in one dimension has not led to any further insights into the workings of SUSY QM. For completeness we will provide in Appendix A a superfield as well as path integral formulation of SUSY quantum mechanics.

In 1983, the concept of a shape invariant potential (SIP) within the structure of SUSY QM was introduced by Gendenshtein. The definition presented was as follows: a potential is said to be shape invariant if its SUSY partner potential has the same spatial dependence as the original potential with possibly altered parameters. It is readily shown that for any SIP, the energy eigenvalue spectra can be obtained algebraically. Much later, a list of SIPs was given and it was shown that the energy eigenfunctions as well as the scattering matrix could also be obtained algebraically for these potentials. It was soon realized that the formalism of SUSY QM plus shape invariance (connected with translations of parameters) was intimately connected to the factorization method of Infeld and Hull.

It is perhaps appropriate at this point to digress a bit and talk about the history of the factorization method. The factorization method was first introduced by Schrödinger to solve the hydrogen atom problem algebraically. Subsequently, Infeld and Hull generalized this method and obtained a wide class of solvable potentials by considering six different forms of factorization. It turns out that the factorization method as well as the methods of SUSY QM including the concept of shape invariance (with translation of parameters), are both reformulations of Riccati's idea of using the equivalence between the solutions of the Riccati equation and a related second order linear differential equation.

The general problem of the classification of SIPs has not yet been solved. A partial classification of the SIPs involving a translation of parameters was done by Cooper, Ginocchio and Khare and will be discussed later in this book. It turns out that in this case one gets all the standard explicitly solvable potentials (those whose energy eigenvalues and wave functions can be explicitly given).

In recent years, one dimensional quantum mechanics has become very important in understanding the exact multi-soliton solutions to certain Hamiltonian dynamical systems governed by high order partial differential equations such as the Korteweg-de Vries and sine-Gordon equations. It was noticed that the solution of these equations was related to solving a quantum mechanics problem whose potential was the solution itself. The technology used to initially find these multi-soliton solutions was based on solving the inverse scattering problem. Since the multi-soliton solutions corresponded to new potentials, it was soon realized that these new solutions were related to potentials which were isospectral to the single soliton potential. Since SUSY QM offers a simple way of obtaining isospectral potentials by using either the Darboux or Abraham-Moses or Pursey techniques, one obtains an interesting connection between the methods of the inverse quantum scattering problem and SUSY QM, and we will discuss this connection. We will also develop new types of approximations to solving quantum mechanics problems that are suggested by several of the topics discussed here, namely the existence of a superpotential, partner potentials, and the hierarchy of Hamiltonians which are isospectral. We will focus on four new approximation methods, the $1/N$ expansion within SUSY QM, δ expansion for the superpotential, a SUSY inspired WKB approximation (SWKB) in quantum mechanics and a variational method which utilizes the hierarchy of Hamiltonians related by SUSY and factorization.

We relegate to Appendix A a discussion of the path integral formulation of SUSY QM. Historically, such a study of SUSY QM was a means of testing ideas for SUSY breaking in quantum field theories. In Appendix B, we briefly discuss the method of operator transformations which allows one to find by coordinate transformations new solvable potentials from old ones. In particular, this allows one to extend the solvable potentials to include the Natanzon class of potentials which are not shape invariant. The new class of solvable potentials have wave functions and energy eigenvalues which are known implicitly rather than explicitly. Perturbative effects on the ground state of a one-dimensional potential are most easily calculated using logarithmic perturbation theory, which is reviewed in Appendix C. Finally, solutions to all the problems are given in Appendix D.

More details and references relevant to this introduction can be found in the review articles and books listed at the end of this chapter.

References

(1) E. Schrödinger, *Further Studies on Solving Eigenvalue Problems by Factorization*, Proc. Roy. Irish Acad. **46A** (1941) 183-206.

(2) L. Infeld and T.E. Hull, *The Factorization Method*, Rev. Mod. Phys. **23** (1951) 21-68.

(3) E. Witten, *Dynamical Breaking of Supersymmetry*, Nucl. Phys. **B188** (1981) 513-554.

(4) F. Cooper and B. Freedman, *Aspects of Supersymmetric Quantum Mechanics*, Ann. Phys. (NY) **146** (1983) 262-288.

(5) D. Lancaster, *Supersymmetry Breakdown in Supersymmetric Quantum Mechanics*, Nuovo Cimento **A79** (1984) 28-44.

(6) L.E. Gendenshtein and I.V. Krive, *Supersymmetry in Quantum Mechanics*, Sov. Phys. Usp. **28** (1985) 645-666.

(7) G. Stedman, *Simple Supersymmetry: Factorization Method in Quantum Mechanics*, Euro. Jour. Phys. **6** (1985) 225-231.

(8) R. Haymaker and A.R.P. Rau, *Supersymmetry in Quantum Mechanics*, Am. Jour. Phys. **54** (1986) 928-936.

(9) R. Dutt, A. Khare and U. Sukhatme, *Supersymmetry, Shape Invariance and Exactly Solvable Potentials*, Am. Jour. Phys. **56** (1988) 163-168.

(10) A. Lahiri, P. Roy and B. Bagchi, *Supersymmetry in Quantum Mechanics*, Int. Jour. Mod. Phys. **A5** (1990) 1383-1456.

(11) O.L. de Lange and R.E. Raab, *Operator Methods in Quantum Mechanics*, Oxford University Press (1991).

(12) F. Cooper, A. Khare and U. Sukhatme, *Supersymmetry and Quantum Mechanics*, Phys. Rep. **251** (1995) 267-385.

(13) G. Junker, *Supersymmetric Methods in Quantum and Statistical Physics*, Springer (1996).

Chapter 2

The Schrödinger Equation in One Dimension

In this book, we are mainly concerned with the quantum mechanical properties of a particle constrained to move along a straight line (the x-axis) under the influence of a time-independent potential $V(x)$. The Hamiltonian H is the sum of a kinetic energy term and a potential energy term, and is given by

$$H = -\frac{\hbar^2}{2m}\frac{d^2}{dx^2} + V(x) . \tag{2.1}$$

We want to obtain solutions of the time independent Schrödinger equation $H\psi = E\psi$, that is

$$-\frac{\hbar^2}{2m}\frac{d^2\psi}{dx^2} + V(x)\psi = E\psi , \tag{2.2}$$

with the wave function $\psi(x)$ constrained to satisfy appropriate boundary conditions.

All elementary quantum mechanics texts discuss piecewise constant potentials with resulting sinusoidally oscillating wave functions in regions where $E > V(x)$, and exponentially damped and growing solutions in regions where $E < V(x)$. The requirements of continuity of ψ and $\psi' \equiv \frac{d\psi}{dx}$ as well as the restrictions coming from the conservation of probability are sufficient to give all the energy eigenstates and scattering properties. Most of the familiar results obtained for piecewise constant potentials are in fact valid for general potentials.

Consider a potential $V(x)$ which goes to a constant value V_{max} at $x \to \pm\infty$, and is less than V_{max} everywhere on the x-axis. A continuous potential of this type with minimum value V_{min} is shown in Fig. 2.1 .

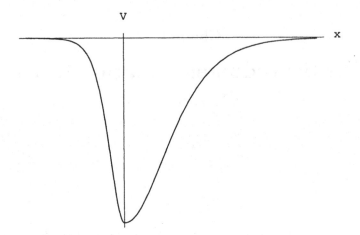

Fig. 2.1   Simple continuous potential with one minimum and equal asymptotes. The potential has both bound states as well as a continuum spectrum.

For $E < V_{min}$, there are no normalizable solutions of eq. (2.2). For $V_{min} < E < V_{max}$, there are discrete values of E for which normalizable solutions exist. These values $E_0, E_1, ...$ are eigenenergies and the corresponding wave functions $\psi_0, \psi_1, ...$ are eigenfunctions. For $E \geq V_{max}$, there is a continuum of energy levels with the wave functions having the behavior $e^{\pm ikx}$ at $x \to \pm\infty$.

In this chapter, we state without proof some general well-known properties of eigenfunctions for both bound state and continuum situations. We will also review the harmonic oscillator problem in the operator formalism in detail, since it is the simplest example of the factorization of a general Hamiltonian discussed in the next chapter. For more details on these subjects, the reader is referred to the references given at the end of this chapter.

2.1   General Properties of Bound States

Discrete bound states exist in the range $V_{min} < E < V_{max}$. The main properties are summarized below:

- The eigenfunctions $\psi_0, \psi_1, ...$ can all be chosen to be real.
- Since the Hamiltonian is Hermitian, the eigenvalues $E_0, E_1, ...$ are necessarily real. Furthermore, for one dimensional problems, the

eigenvalues are non-degenerate.

- The eigenfunctions vanish at $x \to \pm\infty$, and are consequently normalizable: $\int_{-\infty}^{\infty} \psi_i^* \psi_i dx = 1$.
- The eigenfunctions are orthogonal: $\int_{-\infty}^{\infty} \psi_i^* \psi_j dx = 0$, $(i \neq j)$.
- If the eigenstates are ordered according to increasing energy, i.e. $E_0 < E_1 < E_2 < ...$, then the corresponding eigenfunctions are automatically ordered in the number of nodes, with the eigenfunction ψ_n having n nodes.
- ψ_{n+1} has a node located between each pair of consecutive zeros in ψ_n (including the zeros at $x \to \pm\infty$).

2.2 General Properties of Continuum States and Scattering

For $E \geq V_{max}$, there is no quantization of energy. The properties of these continuum states are as follows:

- For any energy E, the wave functions have the behavior $e^{\pm ikx}$ at $x \to \pm\infty$, where $\hbar^2 k^2 / 2m = E - V_{max}$. The quantity k is called the wave number.
- If one considers the standard situation of a plane wave incident from the left, the boundary conditions are

$$\psi_k(x) \quad \to \quad e^{ikx} + R(k)e^{-ikx} \quad , \quad x \to -\infty \; ,$$
$$\psi_k(x) \quad \to \quad T(k)e^{ikx} \quad , \quad x \to \infty \; , \tag{2.3}$$

where $R(k)$ and $T(k)$ are called the reflection and transmission amplitudes (or coefficients) . Conservation of probability guarantees that $|R(k)|^2 + |T(k)|^2 = 1$. For any distinct wave numbers k and k', the wave functions satisfy the orthogonality condition $\int_{-\infty}^{\infty} \psi_k^* \psi_{k'} dx = 0$.

- Considered as functions in the complex k-plane, both $R(k)$ and $T(k)$ have poles on the positive imaginary k-axis which correspond to the bound state eigenvalues of the Hamiltonian.
- The bound state and continuum wave functions taken together form a complete set. An arbitrary function can be expanded as a linear combination of this complete set.

The general properties described above will now be discussed with an explicit example. The potential $V(x) = -12 \operatorname{sech}^2 x$ is an exactly solvable potential discussed in many quantum mechanics texts. It is often called the symmetric Rosen-Morse potential. The eigenstates can be determined either via a traditional treatment of the Schrödinger differential equation by a series method, or, as we shall see a little later in this book, the same results emerge more elegantly from an operator formalism applied to shape invariant potentials. In any case, there are just three discrete eigenstates, given by

$$
\begin{aligned}
E_0 &= -9 \ , \ \psi_0 = \operatorname{sech}^3 x \ , \\
E_1 &= -4 \ , \ \psi_1 = \operatorname{sech}^2 x \tanh x \ , \\
E_2 &= -1 \ , \ \psi_2 = \operatorname{sech} x (5 \tanh^2 x - 1) \ ,
\end{aligned}
\tag{2.4}
$$

with a continuous spectrum for $E \geq 0$. We are using units such that $\hbar = 2m = 1$. Note that ψ_0, ψ_1, ψ_2 have $0, 1, 2$ nodes respectively. The potential has the special property of being reflectionless, that is the reflection coefficient $R(k)$ is zero. The transmission coefficient $T(k)$ is given by

$$
T(k) = \frac{\Gamma(-3 - ik)\Gamma(4 - ik)}{\Gamma(-ik)\Gamma(1 - ik)} \ .
\tag{2.5}
$$

Using the identity $\Gamma(x)\Gamma(1 - x) = \pi / \sin \pi x$, it is easy to check that $|T(k)|^2 = 1$. This result is of course expected from probability conservation. Also, recalling that the Gamma function $\Gamma(x)$ has no zeros and only simple poles at $x = 0, -1, -2, ...$, one sees that in the complex k-plane, the poles of $T(k)$ located on the positive imaginary axis are at $k = 3i, 2i, i$. These poles correspond to the eigenenergies $E_0 = -9, E_1 = -4, E_2 = -1$, since $E = k^2$ with our choice of units.

2.3 The Harmonic Oscillator in the Operator Formalism

The determination of the eigenstates of a particle of mass m in a harmonic oscillator potential $V(x) = \frac{1}{2}kx^2$ is of great physical interest and is discussed in enormous detail in all elementary texts. Defining the angular frequency $\omega \equiv \sqrt{k/m}$, the problem consists of finding all the solutions of the time independent Schrödinger equation

$$
-\frac{\hbar^2}{2m}\frac{d^2\psi}{dx^2} + \frac{1}{2}m\omega^2 x^2 \psi = E\psi \ ,
\tag{2.6}
$$

which satisfy the boundary conditions that $\psi(x)$ vanishes at $x \to \pm\infty$. As is well-known, the solution is a discrete energy spectrum

$$E_n = (n + \frac{1}{2})\hbar\omega , \quad n = 0, 1, 2, \dots ,$$

with corresponding eigenfunctions

$$\psi_n = N_n \exp(-\tilde{x}^2/2) \, H_n(\tilde{x}) , \tag{2.7}$$

where $\tilde{x} = \sqrt{m\omega/\hbar} \, x$, H_n denotes the Hermite polynomial of degree n, and N_n is a normalization constant. The standard procedure for obtaining the eigenstates is to re-scale the Schrödinger equation in terms of dimensionless parameters, determine and factor out the asymptotic behavior, and solve the leftover Hermite differential equation via a series expansion. Imposing boundary conditions leaves only Hermite polynomials as acceptable solutions.

Having gone through the standard solution outlined above, students of quantum mechanics greatly appreciate the elegance and economy of the alternative treatment of the harmonic oscillator potential using raising and lowering operators. We will review this operator treatment in this chapter, since similar ideas of factorizing the Hamiltonian play a crucial role in using supersymmetry to treat general one-dimension potentials.

For the operator treatment, we consider the shifted simple harmonic oscillator Hamiltonian

$$\tilde{H} = -\frac{\hbar^2}{2m}\frac{d^2}{dx^2} + \frac{1}{2}m\omega^2 x^2 - \frac{1}{2}\hbar\omega . \tag{2.8}$$

This shift by a constant energy $\frac{1}{2}\hbar\omega$ is rather trivial, but as we shall see later, is consistent with the standard discussion of unbroken supersymmetry in which the ground state is taken to be at zero energy. Define the raising and lowering operators $a^\dagger$ and a as follows:

$$a^\dagger \equiv \sqrt{\frac{m\omega}{2\hbar}} \left(x - \frac{\hbar}{m\omega}\frac{d}{dx} \right) , \; a \equiv \sqrt{\frac{m\omega}{2\hbar}} \left(x + \frac{\hbar}{m\omega}\frac{d}{dx} \right) . \tag{2.9}$$

It is easy to check that the commutator $[a, a^\dagger]$ is unity, and the shifted harmonic oscillator Hamiltonian is given by

$$\tilde{H} = a^\dagger a \hbar\omega .$$

For any eigenstate $\psi(x)$ of $\tilde{H}$ with eigenvalue $\tilde{E}$, it follows that $a^\dagger\psi$ and $a\psi$ are also eigenstates with eigenvalues $\tilde{E} + \hbar\omega$ and $\tilde{E} - \hbar\omega$ respectively. The proof is straightforward since $[\tilde{H}, a^\dagger] = a^\dagger\hbar\omega$ and $[\tilde{H}, a] = -a\hbar\omega$. Consequently,

$$\tilde{H}a^\dagger\psi = (a^\dagger\tilde{H} - a^\dagger\hbar\omega)\psi = (\tilde{E} + \hbar\omega)a^\dagger\psi ,$$
$$\tilde{H}a\psi = (a\tilde{H} - a\hbar\omega)\psi = (\tilde{E} - \hbar\omega)a\psi . \qquad (2.10)$$

This shows how $a^\dagger$ and a raise and lower the energy eigenvalues. Since $\tilde{H}$ is bounded from below, the lowering process necessarily stops at the ground state $\psi_0(x)$ which is such that $a\psi_0(x) = 0$. This means that the ground state energy of $\tilde{H}$ is zero, and the ground state wave function is given by

$$x\psi_0 + \frac{\hbar}{m\omega}\frac{d\psi_0}{dx} = 0 .$$

This first order differential equation yields the solution

$$\psi_0(x) = N_0 \exp(-m\omega x^2/2\hbar) ,$$

in agreement with eq. (2.7). All higher eigenstates are obtained via application of the raising operator $a^\dagger$:

$$\psi_n = N_n(a^\dagger)^n\psi_0 , \quad \tilde{E}_n = n\hbar\omega , \quad (n = 0, 1, 2, \ldots) . \qquad (2.11)$$

Clearly the simple harmonic oscillator Hamiltonian H has the same eigenfunctions ψ_n, but the corresponding eigenvalues are $E_n = (n + \frac{1}{2})\hbar\omega$, $(n = 0, 1, 2, \ldots)$.

References

(1) L.D. Landau and E.M Lifshitz, *Quantum Mechanics*, Pergamon Press (1958).
(2) A. Messiah, *Quantum Mechanics*, North-Holland (1958).
(3) J. Powell and B. Crasemann, *Quantum Mechanics*, Addison-Wesley (1961).

Problems

1. Consider the infinite square well potential with $V(x) = 0$ for $0 < x < L$ and $V(x) = \infty$ outside the well. This is usually the first potential solved in quantum mechanics courses! Show that there are an infinite number of discrete bound states with eigenenergies $E_n = (n + 1)^2 h^2 / 8mL^2$, ($n = 0, 1, 2, 3, ...$), and obtain the corresponding normalized eigenfunctions. Show that the eigenfunctions corresponding to different energies are orthogonal. Compute the locations of the zeros of ψ_{n+1} and ψ_n, and verify that ψ_{n+1} has exactly one zero between consecutive zeros of ψ_n. The eigenfunctions are sketched in Fig. 3.2.

2. Consider a one dimensional potential well given by $V = 0$ in region I $[0 < x < a/2]$, $V = V_0$ in region II $[a/2 < x < a]$, and $V = \infty$ for $x < 0$, $x > a$. We wish to study the eigenstates of this potential as the strength V_0 is varied from zero to infinity.

(i) What are the eigenvalues E_n for the limiting cases $V_0 = 0$ and $V_0 = \infty$? Measure all energies in terms of the natural energy unit $\hbar^2 \pi^2 / 2ma^2$ for this problem.

(ii) For a general value of V_0, write down the wave functions in region I and region II, and obtain the transcendental equation which gives the eigenenergies. [Note that some of the eigenenergies may be less than V_0].

(iii) Solve the transcendental equations obtained in part (ii) numerically to determine the two lowest eigenenergies E_0 and E_1 for several choices of V_0. Plot E_0 and E_1 as functions of V_0.

(iv) Find the critical value V_{0C} for which $E_0 = V_{0C}$, and carefully plot the ground state eigenfunction $\psi_0(x)$ for this special situation.

3. Using the explicit expressions for the raising operator $a^\dagger$ and the ground state wave function $\psi_0(x)$, compute the excited state wave functions $\psi_1(x)$, $\psi_2(x)$ and $\psi_3(x)$ for a harmonic oscillator potential. Locate the zeros, and verify that $\psi_{n+1}(x)$ has a node between each pair of successive nodes of $\psi_n(x)$ for $n = 0, 1, 2$.

4. Consider the one-dimensional harmonic oscillator potential. Using the Heisenberg equations of motion for x and p, find the time dependence of a and $a^\dagger$ and hence work out the unequal time commutators $[x(t), x(t')]$,

$[p(t), p(t')]$, $[x(t), p(t')]$.

5. Suppose instead of the Bose oscillator, one had a Fermi oscillator i.e. where a and $a^\dagger$ at equal time satisfy the anti-commutation relations

$$\{\, a\, ,\, a\,\} = 0,\, \{\, a^\dagger\, ,\, a^\dagger\, \} = 0,\, \{\, a\, ,\, a^\dagger\, \} = 1\ .$$

Using $H = (1/2)(aa^\dagger - a^\dagger a)\hbar\omega$, work out the eigenvalues of the number operator and hence those of H.

Chapter 3

Factorization of a General Hamiltonian

Starting from a single particle quantum mechanical Hamiltonian

$$H_1 \equiv -\frac{\hbar^2}{2m}\frac{d^2}{dx^2} + V_1(x) \ ,$$

in principle, all the bound state and scattering properties can be calculated. Instead of starting from a given potential $V_1(x)$, one can equally well start by specifying the ground state wave function $\psi_0(x)$ which is nodeless and vanishes at $x = \pm\infty$. It is often not appreciated that once one knows the ground state wave function, then one knows the potential (up to a constant). Without loss of generality, we can choose the ground state energy $E_0^{(1)}$ of H_1 to be zero. Then the Schrödinger equation for the ground state wave function $\psi_0(x)$ is

$$-\frac{\hbar^2}{2m}\frac{d^2\psi_0}{dx^2} + V_1(x)\psi_0(x) = 0 \ , \tag{3.1}$$

so that

$$V_1(x) = \frac{\hbar^2}{2m}\frac{\psi_0''(x)}{\psi_0(x)} \ . \tag{3.2}$$

This allows a determination of the potential $V_1(x)$ from a knowledge of its ground state wave function. It is now easy to factorize the Hamiltonian as follows:

$$H_1 = A^\dagger A \ ,$$

15

where

$$A = \frac{\hbar}{\sqrt{2m}} \frac{d}{dx} + W(x) \; , \; A^\dagger = \frac{-\hbar}{\sqrt{2m}} \frac{d}{dx} + W(x) \; . \tag{3.4}$$

This allows us to identify

$$V_1(x) = W^2(x) - \frac{\hbar}{\sqrt{2m}} W'(x) \; , \tag{3.5}$$

which is the well-known Riccati equation. The quantity $W(x)$ is generally referred to as the superpotential in SUSY QM literature. The solution for $W(x)$ in terms of the ground state wave function is

$$W(x) = -\frac{\hbar}{\sqrt{2m}} \frac{\psi_0'(x)}{\psi_0(x)} \; . \tag{3.6}$$

This solution is obtained by recognizing that once we satisfy $A\psi_0 = 0$, we automatically have a solution to $H_1\psi_0 = A^\dagger A\psi_0 = 0$.

The next step in constructing the SUSY theory related to the original Hamiltonian H_1 is to define the operator $H_2 = AA^\dagger$ obtained by reversing the order of A and $A^\dagger$. A little simplification shows that the operator H_2 is in fact a Hamiltonian corresponding to a new potential $V_2(x)$:

$$H_2 = -\frac{\hbar^2}{2m} \frac{d^2}{dx^2} + V_2(x) \; , \; V_2(x) = W^2(x) + \frac{\hbar}{\sqrt{2m}} W'(x) \; . \tag{3.7}$$

The potentials $V_1(x)$ and $V_2(x)$ are known as supersymmetric partner potentials.

As we shall see, the energy eigenvalues, the wave functions and the S-matrices of H_1 and H_2 are related. To that end notice that the energy eigenvalues of both H_1 and H_2 are positive semi-definite ($E_n^{(1,2)} \geq 0$). For $n > 0$, the Schrödinger equation for H_1

$$H_1\psi_n^{(1)} = A^\dagger A\psi_n^{(1)} = E_n^{(1)}\psi_n^{(1)} \tag{3.8}$$

implies

$$H_2(A\psi_n^{(1)}) = AA^\dagger A\psi_n^{(1)} = E_n^{(1)}(A\psi_n^{(1)}) \; . \tag{3.9}$$

Similarly, the Schrödinger equation for H_2

$$H_2\psi_n^{(2)} = AA^\dagger\psi_n^{(2)} = E_n^{(2)}\psi_n^{(2)} \tag{3.10}$$

mplies

$$H_1(A^\dagger \psi_n^{(2)}) = A^\dagger A A^\dagger \psi_n^{(2)} = E_n^{(2)}(A^\dagger \psi_n^{(2)}) . \tag{3.11}$$

From eqs. (3.8)-(3.11) and the fact that $E_0^{(1)} = 0$, it is clear that the eigenvalues and eigenfunctions of the two Hamiltonians H_1 and H_2 are related by $(n = 0, 1, 2, ...)$

$$E_n^{(2)} = E_{n+1}^{(1)}, \qquad E_0^{(1)} = 0 , \tag{3.12}$$

$$\psi_n^{(2)} = [E_{n+1}^{(1)}]^{-1/2} A \psi_{n+1}^{(1)} , \tag{3.13}$$

$$\psi_{n+1}^{(1)} = [E_n^{(2)}]^{-1/2} A^\dagger \psi_n^{(2)} . \tag{3.14}$$

Notice that if $\psi_{n+1}^{(1)}$ ($\psi_n^{(2)}$) of H_1 (H_2) is normalized then the wave function $\psi_n^{(2)}$ ($\psi_{n+1}^{(1)}$) in eqs. (3.13) and (3.14) is also normalized. Further, the operator $A(A^\dagger)$ not only converts an eigenfunction of $H_1(H_2)$ into an eigenfunction of $H_2(H_1)$ with the same energy, but it also destroys (creates) an extra node in the eigenfunction. Since the ground state wave function of H_1 is annihilated by the operator A, this state has no SUSY partner. Thus the picture we get is that knowing all the eigenfunctions of H_1 we can determine the eigenfunctions of H_2 using the operator A, and vice versa using $A^\dagger$ we can reconstruct all the eigenfunctions of H_1 from those of H_2 except for the ground state. This is illustrated in Fig. 3.1 .

The underlying reason for the degeneracy of the spectra of H_1 and H_2 can be understood most easily from the properties of the SUSY algebra. That is we can consider a matrix SUSY Hamiltonian of the form

$$H = \begin{bmatrix} H_1 & 0 \\ 0 & H_2 \end{bmatrix} , \tag{3.15}$$

which contains both H_1 and H_2. This matrix Hamiltonian is part of a closed algebra which contains both bosonic and fermionic operators with commutation and anti-commutation relations. We consider the operators

$$Q = \begin{bmatrix} 0 & 0 \\ A & 0 \end{bmatrix} , \tag{3.16}$$

$$Q^\dagger = \begin{bmatrix} 0 & A^\dagger \\ 0 & 0 \end{bmatrix} , \tag{3.17}$$

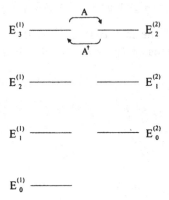

Fig. 3.1   Energy levels of two (unbroken) supersymmetric partner potentials. The action of the operators A and $A^\dagger$ are displayed. The levels are degenerate except that V_1 has an extra state at zero energy.

in conjunction with H. The following commutation and anticommutation relations then describe the closed superalgebra $sl(1/1)$:

$$[H,Q] = [H,Q^\dagger] = 0 \ ,$$
$$\{Q,Q^\dagger\} = H \ , \quad \{Q,Q\} = \{Q^\dagger,Q^\dagger\} = 0 \ . \tag{3.18}$$

The fact that the supercharges Q and $Q^\dagger$ commute with H is responsible for the degeneracy in the spectra of H_1 and H_2. The operators Q and $Q^\dagger$ can be interpreted as operators which change bosonic degrees of freedom into fermionic ones and vice versa. This will be elaborated further below using the example of the SUSY harmonic oscillator. There are various ways of classifying SUSY QM algebras in the literature. One way is by counting the number of anticommuting Hermitian generators $Q_i, i = 1, \cdots, N$ so that an N extended supersymmetry algebra would have

$$\{Q_i, Q_j\} = H\delta_{ij} \ ; \ Q_i = Q_i^\dagger ; \ [H, Q_i] = 0 \ ; \ H = \frac{2}{N}\sum_{i=1}^{N} Q_i^2 \ . \tag{3.19}$$

When $N = 2M$, we can define complex supercharges:

$$\tilde{Q}_i = \frac{Q_{2i-1} + iQ_{2i}}{\sqrt{2}} \ .$$

The usual SUSY would be an $N = 2$ SUSY algebra, with

$$Q = \frac{Q_1 + iQ_2}{\sqrt{2}} \ .$$

Summarizing, we have seen that if there is an exactly solvable potential with at least one bound state, then we can always construct its SUSY partner potential and it is also exactly solvable. In particular, its bound state energy eigenstates are easily obtained by using eq. (3.13).

Let us look at a well known potential, namely the infinite square well and determine its SUSY partner potential. Consider a particle of mass m in an infinite square well potential of width L:

$$\begin{aligned} V(x) &= 0, & 0 \leq x \leq L, \\ &= \infty, & -\infty < x < 0, x > L. \end{aligned} \qquad (3.20)$$

The normalized ground state wave function is known to be

$$\psi_0^{(1)} = (2/L)^{1/2} \sin(\pi x/L), \qquad 0 \leq x \leq L, \qquad (3.21)$$

and the ground state energy is

$$E_0 = \frac{\hbar^2 \pi^2}{2mL^2} \ .$$

Subtracting off the ground state energy so that the Hamiltonian can be factorized, we have for $H_1 = H - E_0$ that the energy eigenvalues are

$$E_n^{(1)} = \frac{n(n+2)}{2mL^2} \hbar^2 \pi^2, \ n = 0, 1, 2, \ldots, \qquad (3.22)$$

and the normalized eigenfunctions are

$$\psi_n^{(1)} = (2/L)^{1/2} \sin \frac{(n+1)\pi x}{L}, \qquad 0 \leq x \leq L. \qquad (3.23)$$

The superpotential for this problem is readily obtained using eq. (3.6)

$$W(x) = -\frac{\hbar}{\sqrt{2m}} \frac{\pi}{L} \cot(\pi x/L), \qquad (3.24)$$

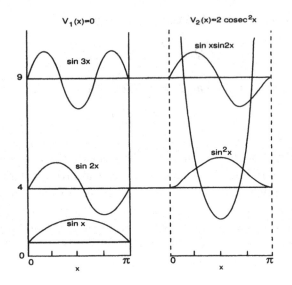

Fig. 3.2   The infinite square well potential $V = 0$ of width π and its partner potential $V = 2 \operatorname{cosec}^2 x$ in units $\hbar = 2m = 1$

and hence the supersymmetric partner potential V_2 is

$$V_2(x) = \frac{\hbar^2 \pi^2}{2mL^2} [2 \operatorname{cosec}^2(\pi x/L) - 1] . \tag{3.25}$$

The wave functions for H_2 are obtained by applying the operator A to the wave functions of H_1. In particular we find that the normalized ground and first excited state wave functions are

$$\psi_0^{(2)} = -2\sqrt{\frac{2}{3L}} \sin^2(\pi x/L) , \quad \psi_1^{(2)} = -\frac{2}{\sqrt{L}} \sin(\pi x/L) \sin(2\pi x/L) . \tag{3.26}$$

Thus we have shown using SUSY that two rather different potentials corresponding to H_1 and H_2 have exactly the same spectra except for the fact that H_2 has one fewer bound state. In Fig. 3.2 we show the supersymmetric partner potentials V_1 and V_2 and the first few eigenfunctions. For convenience we have chosen $L = \pi$ and $\hbar = 2m = 1$.

Supersymmetry also allows one to relate the reflection and transmission coefficients in situations where the two partner potentials have continuous spectra. In order for scattering to take place in both of the partner potentials, it is necessary that the potentials $V_{1,2}$ are finite as $x \to -\infty$ or as

$x \to +\infty$ or both. Let us define

$$W(x \to \pm\infty) \equiv W_{\pm} . \qquad (3.27)$$

Then it follows that

$$V_{1,2} \to W_{\pm}^2 \qquad \text{as } x \to \pm\infty . \qquad (3.28)$$

Let us consider an incident plane wave e^{ikx} of energy E coming from the direction $x \to -\infty$. As a result of scattering from the potentials $V_{1,2}(x)$ one would obtain transmitted waves $T_{1,2}(k)e^{ik'x}$ and reflected waves $R_{1,2}(k)e^{-ikx}$. Thus we have

$$\psi^{(1,2)}(k, x \to -\infty) \to e^{ikx} + R_{1,2}e^{-ikx} ,$$
$$\psi^{(1,2)}(k', x \to +\infty) \to T_{1,2}e^{ik'x} , \qquad (3.29)$$

where k and k' are given by

$$k = (E - W_-^2)^{1/2} , \qquad k' = (E - W_+^2)^{1/2} . \qquad (3.30)$$

SUSY connects continuum wave functions of H_1 and H_2 having the same energy analogously to what happens in the discrete spectrum. Thus using eqs. (3.13) and (3.14) we have the relationships:

$$e^{ikx} + R_1 e^{-ikx} = N[(-ik + W_-)e^{ikx} + (ik + W_-)e^{-ikx}R_2] ,$$
$$T_1 e^{ik'x} = N[(-ik' + W_+)e^{ik'x}T_2] , \qquad (3.31)$$

where N is an overall normalization constant. On equating terms with the same exponent and eliminating N, we find:

$$R_1(k) = \left(\frac{W_- + ik}{W_- - ik}\right) R_2(k) ,$$
$$T_1(k) = \left(\frac{W_+ - ik'}{W_- - ik}\right) T_2(k) . \qquad (3.32)$$

A few remarks are in order at this stage.
1) Clearly $|R_1|^2 = |R_2|^2$ and $|T_1|^2 = |T_2|^2$, that is the partner potentials have identical reflection and transmission probabilities.
2) $R_1(T_1)$ and $R_2(T_2)$ have the same poles in the complex plane except that $R_1(T_1)$ has an extra pole at $k = -iW_-$. This pole is on the positive imaginary axis only if $W_- < 0$ in which case it corresponds to a zero energy bound state.

(3) For the special case $W_+ = W_-$, we have $T_1(k) = T_2(k)$.

(4) When $W_- = 0$, then $R_1(k) = -R_2(k)$.

It is clear from these remarks that if one of the partner potentials is a constant potential (i.e. a free particle), then the other partner will be of necessity reflectionless. In this way we can understand the reflectionless potentials of the form $V(x) = A \operatorname{sech}^2 \alpha x$ which play a critical role in understanding the soliton solutions of the Korteweg-de Vries (KdV) hierarchy which we will discuss later. Let us consider the superpotential

$$W(x) = A \tanh \alpha x . \tag{3.33}$$

The two partner potentials are

$$V_1 = A^2 - A\,(A + \alpha \frac{\hbar}{\sqrt{2m}})\operatorname{sech}^2 \alpha x ,$$

$$V_2 = A^2 - A(A - \alpha \frac{\hbar}{\sqrt{2m}})\operatorname{sech}^2 \alpha x .$$

For the choice $A = \alpha \frac{\hbar}{\sqrt{2m}}$, $V_2(x)$ corresponds to a constant potential and hence the corresponding V_1 is a reflectionless potential. It is worth noting that V_1 is $\hbar$-dependent. One can in fact rigorously show, though it is not mentioned in most textbooks, that the reflectionless potentials are necessarily $\hbar$-dependent.

So far we have discussed SUSY QM on the full line ($-\infty \le x \le \infty$). Many of these results have analogs for the n-dimensional potentials with spherical symmetry. For example, for spherically symmetric potentials in three dimensions one can make a partial wave expansion in terms of the wave functions:

$$\psi_{nlm}(r,\theta,\phi) = \frac{1}{r} R_{nl}(r) Y_{lm}(\theta,\phi) . \tag{3.35}$$

Then it is easily shown that the reduced radial wave function R_{nl} satisfies the one-dimensional Schrödinger equation ($0 \le r \le \infty$)

$$-\frac{\hbar^2}{2m}\frac{d^2 R_{nl}(r)}{dr^2} + \left[V(r) + \frac{l(l+1)\hbar^2}{2mr^2} \right] R_{nl}(r) = E R_{nl}(r) . \tag{3.36}$$

We notice that this is a Schrödinger equation for an effective one dimensional potential which contains the original potential plus an angular momentum barrier. The asymptotic form of the radial wave function for the

l'th partial wave is

$$R(r,l) \rightarrow \frac{1}{2k'}[S^l(k')e^{ik'r} - (-1)^l e^{-ik'r}] , \qquad (3.37)$$

where S^l is the scattering function for the l'th partial wave, i.e. $S^l(k) = e^{i\delta_l(k)}$ and δ is the phase shift.

For this case we find the relations:

$$S_1^l(k') = \left(\frac{W_+ - ik'}{W_+ + ik'}\right) S_2^l(k') . \qquad (3.38)$$

Here $W_+ = W(r \rightarrow \infty)$. Note that, in this case, W and the potential are related by

$$W^2(r) - \frac{\hbar}{\sqrt{2m}} W'(r) = V(r) + \frac{l(l+1)\hbar^2}{2mr^2} - E_0^{(l)} . \qquad (3.39)$$

3.1 Broken Supersymmetry

We have seen that when the ground state wave function of H_1 is known, then we can factorize the Hamiltonian and find a SUSY partner Hamiltonian H_2. Now let us consider the converse problem. Suppose we are given a superpotential $W(x)$. In this case there are two possibilities. The candidate ground state wave function is the ground state for H_1 or H_2 and can be obtained from:

$$A\psi_0^{(1)}(x) = 0 \qquad \Rightarrow \psi_0^{(1)}(x) = N \exp\left(-\frac{\sqrt{2m}}{\hbar} \int^x W(y) \, dy\right) , \qquad (3.40)$$

$$A^\dagger \psi_0^{(2)}(x) = 0 \qquad \Rightarrow \psi_0^{(2)}(x) = N \exp\left(+\frac{\sqrt{2m}}{\hbar} \int^x W(y) \, dy\right) . \qquad (3.41)$$

By convention, we shall always choose W in such a way that amongst H_1, H_2 only H_1 (if at all) will have a normalizable zero energy ground state eigenfunction. This is ensured by choosing W such that $W(x)$ is positive(negative) for large positive(negative) x. This defines H_1 to have fermion number zero in our later formal treatment of SUSY.

If there are no normalizable solutions of this form, then H_1 does not have a zero eigenvalue and SUSY is broken. Let us now be more precise. A symmetry of the Hamiltonian (or Lagrangian) can be spontaneously broken if the lowest energy solution does not respect that symmetry, as for example in a ferromagnet, where rotational invariance of the Hamiltonian is broken by the ground state. We can define the ground state in our system by a two dimensional column vector:

$$|0> \equiv \psi_0(x) = \begin{bmatrix} \psi_0^{(1)}(x) \\ \psi_0^{(2)}(x) \end{bmatrix} . \tag{3.42}$$

For SUSY to be unbroken requires

$$Q|0> = Q^\dagger|0> = 0|0> . \tag{3.43}$$

Thus we have immediately from eq. (3.18) that the ground state energy must be zero in this case. For all the cases we discussed previously, the ground state energy was indeed zero and hence the ground state wave function for the matrix Hamiltonian can be written:

$$\psi_0(x) = \begin{bmatrix} \psi_0^{(1)}(x) \\ 0 \end{bmatrix} , \tag{3.44}$$

where $\psi_0^{(1)}(x)$ is given by eq. (3.40).

If we consider superpotentials of the form

$$W(x) = gx^n , \tag{3.45}$$

then for n odd and g positive one always has a normalizable ground state wave function (this is also true for g negative since in that case we can choose $W(x) = -gx^n$). However for the case n even and g arbitrary, there is no candidate matrix ground state wave function that is normalizable. In this case the potentials V_1 and V_2 have degenerate positive ground state energies and neither Q nor $Q^\dagger$ annihilate the matrix ground state wave function as given by eq. (3.42).

Thus we have the immediate result that if the ground state energy of the matrix Hamiltonian is non-zero then SUSY is broken. For the case of broken SUSY the operators A and $A^\dagger$ no longer change the number of nodes and there is a 1-1 pairing of all the eigenstates of H_1 and H_2. The precise relations that one now obtains are:

$$E_n^{(2)} = E_n^{(1)} > 0, \quad n = 0, 1, 2, \dots \tag{3.46}$$

$$\psi_n^{(2)} = [E_n^{(1)}]^{-1/2} A \psi_n^{(1)} , \tag{3.47}$$

$$\psi_n^{(1)} = [E_n^{(2)}]^{-1/2} A^\dagger \psi_n^{(2)} , \tag{3.48}$$

while the relationship between the scattering amplitudes is still given by eqs. (3.32) or (3.38). The breaking of SUSY can be described by a topological quantum number called the Witten index which we will discuss later. Let us however remember that in general if the sign of $W(x)$ is opposite as we approach infinity from the positive and the negative sides, then SUSY is unbroken, whereas in the other case it is always broken.

Given any nonsingular potential $\tilde{V}(x)$ with eigenfunctions $\psi_n(x)$ and eigenvalues E_n ($n = 0, 1, 2, ...$), let us now enquire how one can find the most general superpotential $W(x)$ which will give $\tilde{V}(x)$ up to an additive constant. To answer this question consider the Schrödinger equation for $\tilde{V}(x)$:

$$-\phi'' + \tilde{V}(x)\phi = \epsilon\phi , \tag{3.49}$$

where ϵ is a constant energy to be chosen later. For convenience, and without loss of generality, we will always choose a solution $\phi(x)$ of eq. (3.49) which vanishes at $x = -\infty$. Note that whenever ϵ corresponds to one of the eigenvalues E_n, the solution $\phi(x)$ is the eigenfunction $\psi_n(x)$. If one defines the quantity $W_\phi = -\phi'/\phi$ and takes it to be the superpotential, then clearly the partner potentials generated by W_ϕ are

$$V_{2(\phi)} = W_\phi^2 + W_\phi' , \quad V_{1(\phi)} = W_\phi^2 - W_\phi' = \frac{\phi''}{\phi} = \tilde{V}(x) - \epsilon , \tag{3.50}$$

where we have used eq. (3.49) for the last step. The eigenvalues of $V_{1(\phi)}$ are therefore given by

$$E_{n(\phi)} = E_n - \epsilon . \tag{3.51}$$

One usually takes ϵ to be the ground state energy E_0 and ϕ to be the ground state wave function $\psi_0(x)$, which makes $E_{0(\phi)} = 0$ and gives the familiar case of unbroken SUSY. With this choice, the superpotential

$$W_0(x) = -\psi_0'/\psi_0$$

is nonsingular, since $\psi_0(x)$ is normalizable and has no nodes. The partner potential $V_{2(\phi)}$ has no eigenstate at zero energy since $A_0\psi_0(x) = [d/dx + W_0(x)]\psi_0(x) = 0$; however, the remaining eigenvalues of $V_{2(\phi)}$ are degenerate with those of $V_{1(\phi)}$.

Let us now consider what happens for other choices of ϵ, both below and above the ground state energy E_0. For $\epsilon < E_0$, the solution $\phi(x)$ has no nodes, and has the same sign for the entire range $-\infty < x < +\infty$. The corresponding superpotential $W_\phi(x)$ is nonsingular. Hence the eigenvalue spectra of $V_{1(\phi)}$ and $V_{2(\phi)}$ are completely degenerate and the energy eigenvalues are given by eq. (3.51). In particular, $E_{0(\phi)} = E_0 - \epsilon$ is positive. Here, W_ϕ has the same sign at $x = \pm\infty$, and we have the case of broken SUSY. For the case when ϵ is above E_0, the solution $\phi(x)$ has one or more nodes, at which points the superpotential $W(x)$ and consequently the supersymmetric partner potential $V_{2(\phi)}$ is singular. Although singular potentials have been discussed in the literature, we will not pursue this topic further here.

As discussed earlier, for SUSY to be a good symmetry, the operators Q and $Q^\dagger$ must annihilate the vacuum. Thus the ground state energy of the super-Hamiltonian must be zero since

$$H = \{Q^\dagger, Q\} \ .$$

Witten proposed an index to determine whether SUSY is broken in supersymmetric field theories. The Witten index is defined by

$$\Delta = \mathrm{Tr}(-1)^F \ , \tag{3.52}$$

where the trace is over all the bound states and continuum states of the super-Hamiltonian. For SUSY QM, the fermion number $n_F \equiv F$ is defined by $\frac{1}{2}(1 - \sigma_3)$ and we can represent $(-1)^F$ by the Pauli matrix σ_3. If we write the eigenstates of H as the vector:

$$\psi_n(x) = \begin{bmatrix} \psi_n^{(+)}(x) \\ \psi_n^{(-)}(x) \end{bmatrix} \ , \tag{3.53}$$

then the $\pm$ corresponds to the eigenvalues of $(-1)^F$ being ± 1. For our conventions the eigenvalue $+1$ corresponds to H_1 and the eigenvalue -1 corresponds to H_2. Since the bound states of H_1 and H_2 are paired, except for the case of unbroken SUSY where there is an extra state in the bosonic sector with $E = 0$ we expect for the quantum mechanics situation that $\Delta = 0$ for broken SUSY and $\Delta = 1$ for unbroken SUSY. In the general field theory case, Witten gives arguments that in general the index measures $N_+(E = 0) - N_-(E = 0)$. In field theories the Witten index needs to be regulated to be well defined so that one considers instead ($\beta = 1/kT$)

$$\Delta(\beta) = \text{Tr}(-1)^F e^{-\beta H} , \tag{3.54}$$

which for SUSY quantum mechanics becomes

$$\Delta(\beta) = \text{Tr}[e^{-\beta H_1} - e^{-\beta H_2}] . \tag{3.55}$$

After calculating the regulated index one wants to take the limit $\beta \to 0$.

In field theory it is quite hard to determine if SUSY is broken non-perturbatively, and thus SUSY quantum mechanics became a testing ground for different methods to understand non-perturbative SUSY breaking. In the quantum mechanics case, the breakdown of SUSY is related to the question of whether there is a normalizable solution to the equation $Q|0 >= 0|0 >$ which implies

$$\psi_0(x) = N e^{-\int^x W(y)dy} . \tag{3.56}$$

As we said before, if this candidate ground state wave function does not fall off fast enough at $\pm\infty$, then Q does not annihilate the vacuum and SUSY is spontaneously broken. Let us show using a trivial calculation that for two simple polynomial potentials the Witten index does indeed provide the correct answer to the question of SUSY breaking. Let us start from eq. (3.54). We represent $(-1)^F$ by σ_3 and we realize that the limit $\beta \to 0$ corresponds to the classical limit since $T \to \infty$. Thus we can replace the quantum trace by an integration over classical phase space so that

$$\Delta(\beta) = \text{Tr}\sigma_3 \int [\frac{dpdx}{2\pi}] e^{-\beta[p^2/2 + W^2/2 - \sigma_3 W'(x)/2]} . \tag{3.57}$$

Expanding the term proportional to σ_3 in the exponent and taking the trace we obtain

$$\Delta(\beta) = \int [\frac{dpdx}{\pi}] e^{-\beta[p^2/2 + W^2/2]} \sinh(\beta W'(x)/2) . \tag{3.58}$$

We are interested in the regulated index as β tends to 0, so that practically we need to evaluate

$$\Delta(\beta) = \int [\frac{dpdx}{2\pi}] e^{-\beta[p^2/2 + W^2/2]} (\beta W'(x)/2) . \tag{3.59}$$

If we directly evaluate this integral for any potential of the form $W(x) = gx^{2n+1} (g > 0)$, which leads to a normalizable ground state wave function, then all the integrals are gamma functions and we explicitly obtain $\Delta = 1$. If instead $W(x) = gx^{2n}$ so that the candidate ground state wave function is not normalizable then the integrand becomes an odd function of x and therefore vanishes. Thus we see for these simple cases in quantum mechanics that the Witten index coincides with the one obtained by the direct method.

3.2 SUSY Harmonic Oscillator

In Chap. 2 we reviewed the operator treatment of the harmonic oscillator. Here we will first recapitulate those results using scaled variables before generalizing to the SUSY extension of the harmonic oscillator. We will also phrase our discussion in terms of Dirac notation where we talk about state vectors instead of wave functions. We will introduce the Fock space of boson occupation numbers where we label the states by the occupation number n. This means instead of P and q as the basic operators, we instead focus on the creation and annihilation operators a and $a^\dagger$. Using slightly different notation, we rewrite the Hamiltonian for the harmonic oscillator as

$$\mathcal{H} = \frac{P^2}{2m} + \frac{1}{2}m\omega^2 q^2 \ . \tag{3.60}$$

We next rescale the Hamiltonian in terms of dimensionless coordinates and momenta x and p. We put

$$\mathcal{H} = H\hbar\omega \ , \qquad q = (\frac{\hbar}{2m\omega})^{1/2}x \ , \qquad P = (2m\hbar\omega)^{1/2}p \ . \tag{3.61}$$

Then

$$H = p^2 + \frac{x^2}{4} \ , \qquad [x,p] = i \ . \tag{3.62}$$

Now we introduce rescaled creation and annihilation operators by (compare eq. (2.9))

$$a = \frac{x}{2} + ip \ , \qquad a^\dagger = \frac{x}{2} - ip \ . \tag{3.63}$$

Then

$$\begin{aligned} [a,a^\dagger] &= 1 \ , \qquad [N,a] = -a \ , \qquad [N,a^\dagger] = a^\dagger \ , \\ N &= a^\dagger a \ , \qquad H = N + \frac{1}{2} \ . \end{aligned} \tag{3.64}$$

The ground state is defined by

$$a|0 >= 0 , \qquad (3.65)$$

which leads to a first order differential equation for the ground state wave function in the Schrödinger picture. The n particle state (which is the n'th excited wave function in the coordinate representation) is then given by:

$$|n >= \frac{a^{\dagger \, n}}{\sqrt{(n!)}}|0 > . \qquad (3.66)$$

For the case of the SUSY harmonic oscillator one can rewrite the operators Q ($Q^\dagger$) as a product of the bosonic operator a and a fermionic operator ψ. Namely we write $Q = a\psi^\dagger$ and $Q^\dagger = a^\dagger \psi$ where the matrix fermionic creation and annihilation operators are defined via:

$$\psi = \sigma_+ = \begin{pmatrix} 0 & 1 \\ 0 & 0 \end{pmatrix}, \qquad (3.67)$$

$$\psi^\dagger = \sigma_- = \begin{pmatrix} 0 & 0 \\ 1 & 0 \end{pmatrix}. \qquad (3.68)$$

ψ and $\psi^\dagger$ obey the usual algebra of the fermionic creation and annihilation operators discussed in detail in Appendix A, namely, they obey the anti-commutation relations

$$\{\psi^\dagger, \psi\} = 1 , \qquad \{\psi^\dagger, \psi^\dagger\} = \{\psi, \psi\} = 0 , \qquad (3.69)$$

where $\{A, B\} \equiv AB + BA$, as well as obeying the commutation relation

$$[\psi, \psi^\dagger] = \sigma_3 = \begin{pmatrix} 1 & 0 \\ 0 & -1 \end{pmatrix}. \qquad (3.70)$$

The SUSY Hamiltonian can be rewritten in the form

$$H = QQ^\dagger + Q^\dagger Q = (-\frac{d^2}{dx^2} + \frac{x^2}{4})I - \frac{1}{2}[\psi, \psi^\dagger] . \qquad (3.71)$$

The effect of the last term is to remove the zero point energy.

The state vector can be thought of as a matrix in the Schrödinger picture or as the state $|n_b, n_f >$ in this Fock space picture. Since the fermionic creation and annihilation operators obey anti-commutation relations, the fermion number is either zero or one. As stated before, we will choose the ground state of H_1 to have zero fermion number. Then we can introduce the fermion number operator

$$n_F = \frac{1 - \sigma_3}{2} = \frac{1 - [\psi, \psi^\dagger]}{2} . \tag{3.72}$$

Because of the anticommutation relation, n_f can only take on the values 0 and 1. The action of the operators $a, a^\dagger, \psi, \psi^\dagger$ in this Fock space are then:

$$
\begin{aligned}
a|n_b, n_f> &= |n_b - 1, n_f> , \quad \psi|n_b, n_f> = |n_b, n_f - 1> , \\
a^\dagger|n_b, n_f> &= |n_b + 1, n_f> , \quad \psi^\dagger|n_b, n_f> = |n_b, n_f + 1> .
\end{aligned}
\tag{3.73}
$$

We now see that the operator $Q^\dagger = a\psi^\dagger$ has the property of changing a boson into a fermion without changing the energy of the state. This is the boson-fermion degeneracy characteristic of all SUSY theories.

For the general case of SUSY QM, the operator a gets replaced by A in the definition of Q, $Q^\dagger$, i.e. one writes $Q = A\psi^\dagger$ and $Q^\dagger = A^\dagger\psi$. The effect of Q and $Q^\dagger$ are now to relate the wave functions of H_1 and H_2 which have fermion number zero and one respectively but now there is no simple Fock space description in the bosonic sector because the interactions are non-linear. Thus in the general case, we can rewrite the SUSY Hamiltonian in the form

$$H = (-\frac{d^2}{dx^2} + W^2)I - [\psi, \psi^\dagger]W'. \tag{3.74}$$

This form will be useful later when we discuss the Lagrangian formulation of SUSY QM in Appendix A.

3.3 Factorization and the Hierarchy of Hamiltonians

In a previous section we found that once we know the ground state wave function corresponding to a Hamiltonian H_1, we can find the superpotential $W_1(x)$ from eq. (3.6). The resulting operators A_1 and $A_1^\dagger$ obtained from eq. (3.4) can be used to factorize Hamiltonian H_1. We also know that the ground state wave function of the partner Hamiltonian H_2 is determined from the first excited state of H_1 via the application of the operator A_1. This allows a refactorization of the second Hamiltonian in terms of W_2 which can be determined from the ground state wave function of H_2. The partner of this refactorization is now another Hamiltonian H_3. Each of the new Hamiltonians has one fewer bound state, so that this process can be continued until the number of bound states is exhausted. Thus if one has an exactly solvable potential problem for H_1, one can solve for the energy eigenvalues and wave functions for the entire hierarchy of Hamiltonians created by repeated refactorizations. Conversely if we know the ground state wave functions for all the Hamiltonians in this hierarchy, we can reconstruct the solutions of the original problem. Let us now be more specific.

We have seen above that if the ground state energy of a Hamiltonian H_1 is zero then it can always be written in a factorizable form as a product of a pair of linear differential operators. It is then clear that if the ground state energy of a Hamiltonian H_1 is $E_0^{(1)}$ with eigenfunction $\psi_0^{(1)}$ then in view of eq. (3.3), it can always be written in the form (unless stated otherwise, from now on we set $\hbar = 2m = 1$ for simplicity):

$$H_1 = A_1^\dagger A_1 + E_0^{(1)} = -\frac{d^2}{dx^2} + V_1(x) , \qquad (3.75)$$

where

$$A_1 = \frac{d}{dx} + W_1(x) , \quad A_1^\dagger = -\frac{d}{dx} + W_1(x) , \quad W_1(x) = -\frac{d \ln\psi_0^{(1)}}{dx} . \qquad (3.76)$$

The SUSY partner Hamiltonian is then given by

$$H_2 = A_1 A_1^\dagger + E_0^{(1)} = -\frac{d^2}{dx^2} + V_2(x) , \qquad (3.77)$$

where

$$V_2(x) = W_1^2 + W_1' + E_0^{(1)} = V_1(x) + 2W_1' = V_1(x) - 2\frac{d^2}{dx^2}\ln\psi_0^{(1)} . \qquad (3.78)$$

We will introduce the notation that in $E_n^{(m)}$, n denotes the energy level and (m) refers to the m'th Hamiltonian H_m. In view of eqs. (3.12), (3.13) and (3.14) the energy eigenvalues and eigenfunctions of the two Hamiltonians H_1 and H_2 are related by

$$E_{n+1}^{(1)} = E_n^{(2)} , \quad \psi_n^{(2)} = (E_{n+1}^{(1)} - E_0^{(1)})^{-1/2} A_1 \psi_{n+1}^{(1)} . \qquad (3.79)$$

Now starting from H_2 whose ground state energy is $E_0^{(2)} = E_1^{(1)}$ one can similarly generate a third Hamiltonian H_3 as a SUSY partner of H_2 since we can write H_2 in the form:

$$H_2 = A_1 A_1^\dagger + E_0^{(1)} = A_2^\dagger A_2 + E_1^{(1)} , \qquad (3.80)$$

where

$$A_2 = \frac{d}{dx} + W_2(x) , \quad A_2^\dagger = -\frac{d}{dx} + W_2(x) , \quad W_2(x) = -\frac{d \ln\psi_0^{(2)}}{dx} . \qquad (3.81)$$

Continuing in this manner we obtain

$$H_3 = A_2 A_2^\dagger + E_1^{(1)} = -\frac{d^2}{dx^2} + V_3(x) , \qquad (3.82)$$

where

$$\begin{aligned}
V_3(x) &= W_2^2 + W_2' + E_1^{(1)} = V_2(x) - 2\frac{d^2}{dx^2}\ln\psi_0^{(2)} \\
&= V_1(x) - 2\frac{d^2}{dx^2}\ln(\psi_0^{(1)}\psi_0^{(2)}) . \qquad (3.83)
\end{aligned}$$

Furthermore

$$\begin{aligned}
E_n^{(3)} &= E_{n+1}^{(2)} = E_{n+2}^{(1)} , \\
\psi_n^{(3)} &= (E_{n+1}^{(2)} - E_0^{(2)})^{-1/2} A_2 \psi_{n+1}^{(2)} \\
&= (E_{n+2}^{(1)} - E_1^{(1)})^{-1/2}(E_{n+2}^{(1)} - E_0^{(1)})^{-1/2} A_2 A_1 \psi_{n+2}^{(1)} . \qquad (3.84)
\end{aligned}$$

In this way, it is clear that if the original Hamiltonian H_1 has $p(\geq 1)$ bound states with eigenvalues $E_n^{(1)}$, and eigenfunctions $\psi_n^{(1)}$ with $0 \leq n \leq (p-1)$, then we can always generate a hierarchy of $(p-1)$ Hamiltonians $H_2, ...H_p$ such that the m'th member of the hierarchy of Hamiltonians (H_m) has the same eigenvalue spectrum as H_1 except that the first $(m-1)$ eigenvalues of H_1 are missing in H_m. In particular, we can always write $(m = 2, 3, ...p)$:

$$H_m = A_m^\dagger A_m + E_{m-1}^{(1)} = -\frac{d^2}{dx^2} + V_m(x) , \qquad (3.85)$$

where

$$A_m = \frac{d}{dx} + W_m(x) , \quad W_m(x) = -\frac{d \ln\psi_0^{(m)}}{dx} . \qquad (3.86)$$

One also has

$$E_n^{(m)} = E_{n+1}^{(m-1)} = ... = E_{n+m-1}^{(1)} ,$$

$$\psi_n^{(m)} = (E_{n+m-1}^{(1)} - E_{m-2}^{(1)})^{-1/2} ... (E_{n+m-1}^{(1)} - E_0^{(1)})^{-1/2} A_{m-1} ... A_1 \psi_{n+m-1}^{(1)} ,$$

$$V_m(x) = V_1(x) - 2\frac{d^2}{dx^2}\ln(\psi_0^{(1)}...\psi_0^{(m-1)}) . \tag{3.87}$$

In this way, knowing all the eigenvalues and eigenfunctions of H_1 we immediately know all the energy eigenvalues and eigenfunctions of the hierarchy of $p - 1$ Hamiltonians. Further the reflection and transmission coefficients (or phase shifts) for the hierarchy of Hamiltonians can be obtained in terms of R_1, T_1 of the first Hamiltonian H_1 by a repeated use of eq. (3.32). In particular we find

$$R_m(k) = \left(\frac{W_-^{(1)} - ik}{W_-^{(1)} + ik}\right) ... \left(\frac{W_-^{(m-1)} - ik}{W_-^{(m-1)} + ik}\right) R_1(k) ,$$

$$T_m(k) = \left(\frac{W_-^{(1)} - ik}{W_+^{(1)} - ik'}\right) ... \left(\frac{W_-^{(m-1)} - ik}{W_+^{(m-1)} - ik'}\right) T_1(k) , \tag{3.88}$$

where k and k' are given by

$$k = [E - (W_-^{(1)})^2]^{1/2}, \qquad k' = [E - (W_+^{(1)})^2]^{1/2}. \tag{3.89}$$

References

(1) E. Witten, *Dynamical Breaking of Supersymmetry*, Nucl. Phys. **B188** (1981) 513-554.
(2) F. Cooper and B. Freedman, *Aspects of Supersymmetric Quantum Mechanics*, Ann. Phys. **146** (1983) 262-288.
(3) E. Witten, *Constraints on Supersymmetry Breaking*, Nucl. Phys. **B202** (1982) 253-316.
(4) C.V. Sukumar, *Supersymmetry, Factorization of the Schrödinger Equation and a Hamiltonian Hierarchy*, J. Phys. **A18** (1985) L57-L61.

Problems

1. Let $V_1(x)$ denote an infinite square well of width π in the range $0 \leq x \leq \pi$. Compute the potentials $V_m(x)$, $(m = 1, 2, ...)$ in the supersymmetric hierarchy. Show that the energy spectrum of $V_m(x)$ is $E_n^{(m)} = (n+m)^2$, $(n = 0, 1, 2, ...)$. Find explicit expressions for the two lowest lying eigenfunctions $\psi_0^{(m)}$ and $\psi_1^{(m)}$ for $m = 1, 2, 3$.

2. Consider the superpotential $W = ax^3 (a > 0)$. Write down the two partner potentials and plot them as a function of x. Show that one of them is a double well and the other a single well potential.

3. An acceptable ground state wave function on the half line $(0 < r < \infty)$ is $\psi_0(r) = Ar^5 e^{-\beta r}$, since it is nodeless and vanishes at $r = 0, \infty$. Compute and plot the corresponding superpotential $W(r)$ and the supersymmetric partner potentials $V_1(r)$ and $V_2(r)$. Take $\beta = 1$ for making graphs.

4. Consider the superpotential $W(x) = Ax^2 + Bx + C$, where A, B, C are positive constants. Is this an example of broken or unbroken supersymmetry? Taking the values $A = 1/5$, $B = 1$, $C = 0$, compute and plot the partner potentials $V_1(x)$ and $V_2(x)$.

5. Start from the potential $V(x) = -12 \operatorname{sech}^2 x$ $(\hbar = 2m = 1)$ whose eigenspectrum and transmission coefficient have been given in Chap. 2. Work out the corresponding superpotential W and hence the corresponding family of potentials V_2, V_3, V_4. Using the eigenfunctions and transmission coefficient for the potential $V(x)$ given above, obtain the same quantities for the potentials V_2, V_3, V_4.

Chapter 4

Shape Invariance and Solvable Potentials

n Chap. 2 we have reviewed how the one dimensional harmonic oscillator problem can be elegantly solved using the raising and lowering operator method. Using the ideas of SUSY QM developed in Chap. 3 and an integrability condition called the shape invariance condition, we now show hat the operator method for the harmonic oscillator can be generalized o s whole class of shape invariant potentials (SIPs) which includes all the popular, analytically solvable potentials. Indeed, we shall see that for such potentials, the generalized operator method quickly yields all the bound state energy eigenvalues and eigenfunctions as well as the scattering matrix. t turns out that this approach is essentially equivalent to Schrödinger's method of factorization although the language of SUSY is more appealing.

Let us now explain precisely what one means by shape invariance. If he pair of SUSY partner potentials $V_{1,2}(x)$ defined in Chap. 3 are similar n shape and differ only in the parameters that appear in them, then they are said to be shape invariant. More precisely, if the partner potentials $V_{1,2}(x; a_1)$ satisfy the condition

$$V_2(x; a_1) = V_1(x; a_2) + R(a_1), \tag{4.1}$$

where a_1 is a set of parameters, a_2 is a function of a_1 (say $a_2 = f(a_1)$) and the remainder $R(a_1)$ is independent of x, then $V_1(x; a_1)$ and $V_2(x; a_1)$ are said to be shape invariant. The shape invariance condition (4.1) is an integrability condition. Using this condition and the hierarchy of Hamiltonians discussed in Chap. 3 , one can easily obtain the energy eigenvalues and eigenfunctions of any SIP when SUSY is unbroken.

4.1   General Formulas for Bound State Spectrum, Wave Functions and S-Matrix

Let us start from the SUSY partner Hamiltonians H_1 and H_2 whose eigenvalues and eigenfunctions are related by SUSY. Further, since SUSY is unbroken we know that

$$E_0^{(1)}(a_1) = 0, \quad \psi_0^{(1)}(x; a_1) = N \exp\left[-\int^x W_1(y; a_1)dy\right]. \qquad (4.2)$$

We will now show that the entire spectrum of H_1 can be very easily obtained algebraically by using the shape invariance condition (4.1). To that purpose, let us construct a series of Hamiltonians H_s, $s = 1, 2, 3\ldots$. In particular, following the discussion of the last chapter it is clear that if H_1 has p bound states then one can construct p such Hamiltonians $H_1, H_2 \cdots H_p$ and the n'th Hamiltonian H_n will have the same spectrum as H_1 except that the first $n - 1$ levels of H_1 will be absent in H_n. On repeatedly using the shape invariance condition (4.1), it is then clear that

$$H_s = -\frac{d^2}{dx^2} + V_1(x; a_s) + \sum_{k=1}^{s-1} R(a_k), \qquad (4.3)$$

where $a_s = f^{s-1}(a_1)$ i.e. the function f applied $s-1$ times. Let us compare the spectrum of H_s and H_{s+1}. In view of eqs. (4.1) and (4.3) we have

$$H_{s+1} = -\frac{d^2}{dx^2} + V_1(x; a_{s+1}) + \sum_{k=1}^{s} R(a_k)$$

$$= -\frac{d^2}{dx^2} + V_2(x; a_s) + \sum_{k=1}^{s-1} R(a_k). \qquad (4.4)$$

Thus H_s and H_{s+1} are SUSY partner Hamiltonians and hence have identical bound state spectra except for the ground state of H_s whose energy is

$$E_0^{(s)} = \sum_{k=1}^{s-1} R(a_k). \qquad (4.5)$$

This follows from eq. (4.3) and the fact that $E_0^{(1)} = 0$. On going back from H_s to H_{s-1} etc, we would eventually reach H_2 and H_1 whose ground state energy is zero and whose n'th level is coincident with the ground state of the Hamiltonian H_n. Hence the complete eigenvalue spectrum of H_1 is given by

$$E_n^{(1)}(a_1) = \sum_{k=1}^{n} R(a_k); \quad E_0^{(1)}(a_1) = 0. \tag{4.6}$$

We now show that, similar to the case of the one dimensional harmonic oscillator, the bound state wave functions $\psi_n^{(1)}(x; a_1)$ for any shape invariant potential can also be easily obtained from its ground state wave function $\psi_0^{(1)}(x; a_1)$ which in turn is known in terms of the superpotential. This is possible because the operators A and $A^\dagger$ link up the eigenfunctions of the same energy for the SUSY partner Hamiltonians $H_{1,2}$. Let us start from the Hamiltonian H_s as given by eq. (4.3) whose ground state eigenfunction is then given by $\psi_0^{(1)}(x; a_s)$. On going from H_s to H_{s-1} to H_2 to H_1 and using eq. (3.14) we then find that the n'th state unnormalized energy eigenfunction $\psi_n^{(1)}(x; a_1)$ for the original Hamiltonian $H_1(x; a_1)$ is given by

$$\psi_n^{(1)}(x; a_1) \propto A^\dagger(x; a_1) A^\dagger(x; a_2)...A^\dagger(x; a_n)\psi_0^{(1)}(x; a_{n+1}), \tag{4.7}$$

which is clearly a generalization of the operator method of constructing the energy eigenfunctions for the one dimensional harmonic oscillator.

It is often convenient to have explicit expressions for the wave functions. In that case, instead of using the above equation, it is far simpler to use the identity

$$\psi_n^{(1)}(x; a_1) = A^\dagger(x; a_1)\psi_{n-1}^{(1)}(x; a_2). \tag{4.8}$$

Finally, in view of the shape invariance condition (4.1), the relation (3.32) between scattering amplitudes takes a particularly simple form

$$R_1(k; a_1) = \left(\frac{W_-(a_1) + ik}{W_-(a_1) - ik}\right) R_1(k; a_2), \tag{4.9}$$

$$T_1(k; a_1) = \left(\frac{W_+(a_1) - ik'}{W_-(a_1) - ik}\right) T_1(k; a_2), \tag{4.10}$$

thereby relating the reflection and transmission coefficients of the same Hamiltonian H_1 at a_1 and $a_2(= f(a_1))$.

4.2 Strategies for Categorizing Shape Invariant Potentials

Let us now discuss the interesting question of the classification of various solutions to the shape invariance condition (4.1). This is clearly an important problem because once such a classification is available, then one discovers new SIPs which are solvable by purely algebraic methods. Although the general problem is still unsolved, two classes of solutions have been found so far. In the first class, the parameters a_1 and a_2 are related to each other by translation $(a_2 = a_1 + \alpha)$. Remarkably enough, all well known analytically solvable potentials found in most textbooks on nonrelativistic quantum mechanics belong to this class. In the second class, the parameters a_1 and a_2 are related to each other by scaling $(a_2 = qa_1)$.

4.2.1 *Solutions Involving Translation*

We shall now point out the key steps that go into the classification of SIPs in case $a_2 = a_1 + \alpha$. Firstly, one notices the fact that the eigenvalue spectrum of the Schrödinger equation is always such that the n'th eigenvalue E_n for large n obeys the constraint

$$A/n^2 \leq E_n \leq Bn^2 , \qquad (4.11)$$

where the upper bound is saturated by the infinite square well potential while the lower bound is saturated by the Coulomb potential. Thus, for any SIP, the structure of E_n for large n is expected to be of the form

$$E_n \sim \sum_\alpha C_\alpha n^\alpha , \quad -2 \leq \alpha \leq 2 . \qquad (4.12)$$

Now, since for any SIP, E_n is given by eq. (4.6), it follows that if

$$R(a_k) \sim \sum_\beta k^\beta , \qquad (4.13)$$

then

$$-3 \leq \beta \leq 1 . \qquad (4.14)$$

How does one implement this constraint on $R(a_k)$? While one has no rigorous answer to this question, it is easily seen that a fairly general factorizable form of $W(x; a_1)$ which produces the above k-dependence in $R(a_k)$

s given by

$$W(x; a_1) = \sum_{i=1}^{n}[(k_i + c_i)g_i(x) + h_i(x)/(k_i + c_i) + f_i(x)] , \qquad (4.15)$$

where

$$a_1 = (k_1, k_2...) , \quad a_2 = (k_1 + \alpha, k_2 + \beta...) , \qquad (4.16)$$

with c_i, α, β being constants. Note that this ansatz excludes all potentials leading to E_n which contain fractional powers of n. On using the above ansatz for W in the shape invariance condition eq. (4.1), one can obtain the conditions to be satisfied by the functions $g_i(x), h_i(x), f_i(x)$. One important condition is of course that only those superpotentials W are admissible which give a square integrable ground state wave function. The shape invariance condition takes a simple form if we choose a rescaled set of parameters $m = (m_1, m_2, \cdots m_n)$ related by translation by an integer so that

$$V_2(x, m) = V_1(x, m - 1) + R(m - 1) . \qquad (4.17)$$

In terms of the superpotential W one then obtains the differential-difference equation

$$W^2(x, m + 1) - W^2(x, m) + W'(x, m + 1) + W'(x, m) = L(m) - L(m + 1) \qquad (4.18)$$

with $R(m) = L(m) - L(m + 1)$. If we insert the ansatz eq. (4.15) into eq. (4.18), we find that for $n = 2$ there are only two solutions. More precisely, choosing

$$W(x; a_1) = (k_1 + c_1)g_1(x) + (k_2 + c_2)g_2(x) + f_1(x) , \qquad (4.19)$$

we find the two solutions

$$W(x; A, B) = A \tan(\alpha x + x_0) - B \cot(\alpha x + x_0) , \quad A, B > 0 , \quad (4.20)$$

and

$$W(r; A, B) = A \tanh \alpha r - B \coth \alpha r , \quad A > B > 0 , \qquad (4.21)$$

where $0 \le x \le \pi/2\alpha$ and $\psi(x = 0) = \psi(x = \pi/2\alpha) = 0$. For the simplest possibility of $n = 1$, one has a number of solutions to the shape invariance condition (4.1). In Table 4.1, we give expressions for the various shape

Table 4.1   Shape invariant potentials with (n=1,2) in which the parameters a_2 and a_1 are related by translation ($a_2 = a_1 + \beta$). The energy eigenvalues and eigenfunctions are given in units $\hbar = 2m = 1$. The constants A, B, α, ω, l are all taken ≥ 0. Unless otherwise stated, the range of potentials is $-\infty \leq x \leq \infty, 0 \leq r \leq \infty$. For spherically symmetric potentials, the full wave function is $\varphi_{nlm}(r, \theta, \phi) = \varphi_{nl}(r)Y_{lm}(\theta, \phi)$.

Potential	$W(x)$	$V_1(x; a_1)$	a_1
Shifted oscillator	$\frac{1}{2}\omega x - b$	$\frac{1}{4}\omega^2\left(x - \frac{2b}{\omega}\right)^2 - \omega/2$	ω
3-D oscillator	$\frac{1}{2}\omega r - \frac{(l+1)}{r}$	$\frac{1}{4}\omega^2 r^2 + \frac{l(l+1)}{r^2} - (l+3/2)\omega$	l
Coulomb	$\frac{e^2}{2(l+1)} - \frac{(l+1)}{r}$	$-\frac{e^2}{r} + \frac{l(l+1)}{r^2} + \frac{e^4}{4(l+1)^2}$	l
Morse	$A - B\exp(-\alpha x)$	$A^2 + B^2\exp(-2\alpha x)$ $-2B(A + \alpha/2)\exp(-\alpha x)$	A
Scarf II (hyperbolic)	$A\tanh\alpha x + B\,\text{sech}\,\alpha x$	$A^2 + (B^2 - A^2 - A\alpha)\text{sech}^2\alpha x$ $+B(2A + \alpha)\text{sech}\,\alpha x\tanh\alpha x$	A
Rosen-Morse II (hyperbolic)	$A\tanh\alpha x + B/A$ $(B < A^2)$	$A^2 + B^2/A^2 - A(A + \alpha)\text{sech}^2\alpha x$ $+ 2B\tanh\alpha x$	A
Eckart	$-A\coth\alpha r + B/A$ $(B > A^2)$	$A^2 + B^2/A^2 - 2B\coth\alpha r$ $+A(A - \alpha)\text{cosech}^2\alpha r$	A
Scarf I (trigonometric)	$A\tan\alpha x - B\sec\alpha x$ $\left(-\frac{1}{2}\pi \leq \alpha x \leq \frac{1}{2}\pi\right)$	$-A^2 + (A^2 + B^2 - A\alpha)\sec^2\alpha x$ $-B(2A - \alpha)\tan\alpha x\sec\alpha x$	A
Pöschl-Teller	$A\coth\alpha r - B\,\text{cosech}\alpha r$ $(A < B)$	$A^2 + (B^2 + A^2 + A\alpha)\text{cosech}^2\alpha r$ $-B(2A + \alpha)\coth\alpha r\,\text{cosech}\,\alpha r$	A
Rosen-Morse I (trigonometric)	$-A\cot\alpha x - B/A$ $(0 \leq \alpha x \leq \pi)$	$A(A - \alpha)\text{cosec}^2\alpha x + 2B\cot\alpha x$ $-A^2 + B^2/A^2$	A

invariant potentials $V_1(x)$, superpotentials $W(x)$, parameters a_1 and a_2 and the corresponding energy eigenvalues $E_n^{(1)}$. Except for first 3 entries of this table, $W(x + x_0)$ is also a solution. Until recently, these were the only solutions found. However a recent careful study by Cariñena and Ramos of the differential-difference eq. (4.18) has found solutions for arbitrary $n \geq 3$

Note that the wave functions for the first four potentials (Hermite and Laguerre polynomials) are special cases of the confluent hypergeometric function while the rest (Jacobi polynomials) are special cases of the hypergeometric function. Fig. B.1 of Appendix B shows the inter-relations between all the SIPs. In the table $s_1 = s - n + a$, $s_2 = s - n - a$, $s_3 = a - n - s$, $s_4 = -(s + n + a)$.

a_2	Eigenvalue $E_n^{(1)}$	Variable y	Wave function $\psi_n(y)$
ω	$n\omega$	$y = \sqrt{\frac{\omega}{2}}\left(x - \frac{2b}{\omega}\right)$	$\exp\left(-\frac{1}{2}y^2\right) H_n(y)$
$l+1$	$2n\omega$	$y = \frac{1}{2}\omega r^2$	$y^{(l+1)/2} \exp\left(-\frac{1}{2}y\right) L_n^{l+1/2}(y)$
$l+1$	$\frac{e^4}{4(l+1)^2} - \frac{e^4}{4(n+l+1)^2}$	$y = \frac{re^2}{(n+l+1)}$	$y^{(l+1)} \exp\left(-\frac{1}{2}y\right) L_n^{2l+1}(y)$
$A - \alpha$	$A^2 - (A - n\alpha)^2$	$y = \frac{2B}{\alpha}e^{-\alpha x}$, $s = A/\alpha$	$y^{s-n} \exp\left(-\frac{1}{2}y\right) L_n^{2s-2n}(y)$
$A - \alpha$	$A^2 - (A - n\alpha)^2$	$y = \sinh \alpha x$, $s = A/\alpha, \lambda = B/\alpha$	$i^n (1+y^2)^{-s/2} e^{-\lambda \tan^{-1} y}$ $\times P_n^{(i\lambda - s - 1/2, -i\lambda - s - 1/2)}(iy)$
$A - \alpha$	$A^2 - (A - n\alpha)^2$ $-B^2/(A - n\alpha)^2$ $+B^2/A^2$	$y = \tanh \alpha x$, $s = A/\alpha, \lambda = B/\alpha^2$ $a = \lambda/(s - n)$	$(1 - y)^{s_1/2}(1 + y)^{s_2/2}$ $\times P_n^{(s_1, s_2)}(y)$
$A + \alpha$	$A^2 - (A + n\alpha)^2$ $-B^2/(A + n\alpha)^2$ $+B^2/A^2$	$y = \coth \alpha r$, $s = A/\alpha, \lambda = B/\alpha^2$ $a = \lambda/(n + s)$	$(y - 1)^{s_3/2}(y + 1)^{s_4/2}$ $\times P_n^{(s_3, s_4)}(y)$
$A + \alpha$	$(A + n\alpha)^2 - A^2$	$y = \sin \alpha r$, $s = A/\alpha, \lambda = B/\alpha$	$(1 - y)^{(s - \lambda)/2}(1 + y)^{(s + \lambda)/2}$ $\times P_n^{(s - \lambda - 1/2, s + \lambda - 1/2)}(y)$
$A - \alpha$	$A^2 - (A - n\alpha)^2$	$y = \cosh \alpha r$, $s = A/\alpha, \lambda = B/\alpha$	$(y - 1)^{(\lambda - s)/2}(y + 1)^{-(\lambda + s)/2}$ $\times P_n^{(\lambda - s - 1/2, -\lambda - s - 1/2)}(y)$
$A + \alpha$	$(A + n\alpha)^2 - A^2$ $-B^2/(A + n\alpha)^2$ $+B^2/A^2$	$y = i\cot \alpha x$, $s = A/\alpha, \lambda = B/\alpha^2$ $a = \lambda/(s + n)$	$(y^2 - 1)^{-(s + n)/2} \exp(a\alpha x)$ $\times P_n^{(-s - n + ia, -s - n - ia)}(y)$

assuming a solution of the form:

$$W(x, m) = g_0(x) + \sum_{i=1}^{n} m_i g_i(x) . \tag{4.22}$$

Inserting this ansatz into the differential-difference equation, Cariñena and

Ramos find

$$L(m) - L(m+1) = 2 \sum_{j=1}^{n} m_j \left(g'_j + g_j \sum_{i=1}^{n} g_i \right)$$

$$+ \sum_{j=1}^{n} (g'_j + g_j \sum_{i=1}^{n} g_i) + 2 \left(g'_0 + g_0 \sum_{i=1}^{n} g_i \right). \qquad (4.23)$$

Since the coefficients of the powers of each m_i have to be constant, they obtain the following system of first order differential equations to be satisfied,

$$g'_j + g_j \sum_{i=1}^{n} g_i = c_j, \quad \forall j \in \{1, \dots, n\}, \qquad (4.24)$$

$$g'_0 + g_0 \sum_{i=1}^{n} g_i = c_0, \qquad (4.25)$$

where c_i, $i \in \{0, 1, \dots, n\}$ are real constants.

The solution of the system can be found by using barycentric coordinates for the g_i's,

$$g_{cm}(x) = \frac{1}{n} \sum_{i=1}^{n} g_i(x), \qquad (4.26)$$

$$v_j(x) = g_j(x) - g_{cm}(x) = \frac{1}{n} \left(n g_j(x) - \sum_{i=1}^{n} g_i(x) \right), \qquad (4.27)$$

$$c_{cm} = \frac{1}{n} \sum_{i=1}^{n} c_i, \qquad (4.28)$$

where $j \in \{1, \dots, n\}$. Note that not all of the functions v_j are now linearly independent, but only $n-1$ since $\sum_{j=1}^{n} v_j = 0$.

Taking the sum of equations (4.24) one obtains that $n g_{cm}$ satisfies the Riccati equation with constant coefficients

$$n g'_{cm} + (n g_{cm})^2 = n c_{cm}.$$

Using equations (4.27) and (4.24) one finds

$$v'_j = \frac{1}{n} (n g'_j - \sum_{i=1}^{n} g'_i)$$

$$= \frac{1}{n}(g'_j - g'_1 + g'_j - g'_2 + \ldots + g'_j - g'_j + \ldots + g'_j - g'_n)$$
$$= -v_j n g_{cm} + c_j - c_{cm} .$$

The system of equations (4.24) and (4.25) becomes

$$ng'_{cm} + (ng_{cm})^2 = nc_{cm} , \tag{4.29}$$
$$v'_j + v_j n g_{cm} = c_j - c_{cm} , \quad \forall j \in \{2, \ldots, n\} , \tag{4.30}$$
$$g'_0 + g_0 n g_{cm} = c_0 . \tag{4.31}$$

These equations have known solutions and one can therefore reconstruct the superpotential from

$$g_1(x) = g_{cm}(x) - \sum_{i=2}^{n} v_i(x) , \tag{4.32}$$
$$g_j(x) = g_{cm}(x) + v_j(x) , \quad \forall j \in \{2, \ldots, n\} . \tag{4.33}$$

The interested reader is referred to the article by Cariñena and Ramos for details. In general the answer for the superpotential can be given in terms of ratios of sums of sines and cosines or ratios of sums of sinh and cosh. It is suspected, however, that the solutions found for $n \geq 3$ can be mapped into the solutions for $n = 2$ or $n = 1$ by a suitable change of parameters.

Let us now give an example of how the elements of the table are constructed. Consider the superpotential given in eq. (4.20) with $x_0 = 0$. The corresponding partner potentials are

$$V_1(x; A, B) = -(A + B)^2 + A(A - \alpha) \sec^2 \alpha x + B(B - \alpha)\csc^2 \alpha x,$$

$$V_2(x; A, B) = -(A + B)^2 + A(A + \alpha) \sec^2 \alpha x + B(B + \alpha)\csc^2 \alpha x . \tag{4.34}$$

V_1 and V_2 are often called Pöschl-Teller I potentials in the literature. They are shape invariant partner potentials since

$$V_2(x; A, B) = V_1(x; A + \alpha, B + \alpha) + (A + B + 2\alpha)^2 - (A + B)^2 , \tag{4.35}$$

and in this case

$$\{a_1\} = (A, B); \{a_2\} = (A + \alpha, B + \alpha), R(a_1) = (A + B + 2\alpha)^2 - (A + B)^2 . \tag{4.36}$$

In view of eq. (4.6), the bound state energy eigenvalues of the potential $V_1(x; A, B)$ are then given by

$$E_n^{(1)} = \sum_{k=1}^{n} R(a_k) = (A + B + 2n\alpha)^2 - (A + B)^2 . \tag{4.37}$$

The ground state wave function of $V_1(x; A, B)$ is calculated from the superpotential W as given by eq. (4.20). We find

$$\psi_0^{(1)}(x; A, B) \propto (\cos \alpha x)^s (\sin \alpha x)^\lambda , \tag{4.38}$$

where

$$s = A/\alpha ; \quad \lambda = B/\alpha . \tag{4.39}$$

The requirement of $A, B > 0$ that we have assumed in eq. (4.20) guarantees that $\psi_0^{(1)}(x; A, B)$ is well behaved and hence acceptable as $x \longrightarrow 0, \pi/2\alpha$. Using this expression for the ground state wave function and eq. (4.8) one can also obtain explicit expressions for the bound state eigenfunctions $\psi_n^{(1)}(x; A, B)$. In particular, in this case, eq. (4.8) takes the form

$$\psi_n(x; \{a_1\}) = \left(-\frac{d}{dx} + A \tan \alpha x - B \cot \alpha x \right) \psi_{n-1}(x; \{a_2\}) . \tag{4.40}$$

On defining a new variable

$$y = 1 - 2\sin^2 \alpha x , \tag{4.41}$$

and factoring out the ground state state wave function

$$\psi_n(y; \{a_1\}) = \psi_0(y; \{a_1\}) R_n(y; \{a_1\}) , \tag{4.42}$$

with ψ_0 being given by eq. (4.38), we obtain:

$$R_n(y; A, B) = \alpha(1 - y^2)\frac{d}{dy} R_{n-1}(y; A + \alpha, B + \alpha)$$
$$+ [(A - B) - (A + B + \alpha)y] R_{n-1}(y; A + \alpha, B + \alpha) . \tag{4.43}$$

It is then clear that $R_n(y; A, B)$ is proportional to the Jacobi Polynomial $P_n^{a,b}$ so that the unnormalized bound state energy eigenfunctions for this potential are

$$\psi_n(y; A, B) = (1 - y)^{\lambda/2}(1 + y)^{s/2} P_n^{\lambda-1/2, s-1/2}(y) . \tag{4.44}$$

The procedure outlined above has been applied to all known SIPs and the energy eigenfunctions $\psi_n^{(1)}(y)$ have been obtained in Table 4.1, where we also give the variable y for each case.

Several remarks are in order at this time.

(1) The Pöschl-Teller I and II superpotentials as given by eqs. (4.20) and (4.21) respectively have not been included in Table 4.1 since they are equivalent to the Scarf I (trigonometric) and Pöschl-Teller superpotentials

$$
\begin{aligned}
W_1 &= A\tan\alpha x - B\sec\alpha x , \\
W_2 &= A\coth\alpha r - B\operatorname{cosech}\alpha r ,
\end{aligned}
\tag{4.45}
$$

by appropriate redefinition of the parameters. For example, one can write

$$
W_2 = (\frac{A+B}{2})\tanh(\frac{\alpha r}{2}) - (\frac{B-A}{2})\coth(\frac{\alpha r}{2}) ,
\tag{4.46}
$$

which is just the Pöschl-Teller II superpotential of eq. (4.21) with redefined parameters.

(2) Throughout this section we have used the convention of $\hbar = 2m = 1$. It would naively appear that if we had not put $\hbar = 1$, then the shape invariant potentials as given in Table 4.1 would all be $\hbar$ dependent. However, it is worth noting that in each and every case, the $\hbar$ dependence is only in the constant multiplying the x-dependent function so that in each case we can always redefine the constant multiplying the function and obtain an $\hbar$ independent potential. For example, corresponding to the superpotential given by eq. (4.20), the $\hbar$ dependent potential is given by ($2m = 1$)

$$
\begin{aligned}
V_1(x; A, B) = W^2 - \hbar W' &= -(A+B)^2 + A(A+\hbar\alpha)\sec^2\alpha x \\
&+ B(B+\hbar\alpha)\operatorname{cosec}^2\alpha x .
\end{aligned}
\tag{4.47}
$$

On redefining

$$
A(A+\hbar\alpha) = a ; \quad B(B+\hbar\alpha) = b ,
\tag{4.48}
$$

where a, b are $\hbar$ independent parameters, we then have an $\hbar$ independent potential.

(3) In Table 4.1, we have given conditions (like $A > 0$, $B > 0$) for the superpotential (4.20), so that $\psi_0^{(1)} = N \exp\left(-\int^x W(y)dy\right)$ is an acceptable ground state energy eigenfunction. Instead one can also write down conditions for $\psi_0^{(2)} = N \exp\left(\int^x W(y)dy\right)$ to be an acceptable ground state energy eigenfunction.

(4) It may be noted that the Coulomb as well as the harmonic oscillator potentials in n-dimensions are also shape invariant potentials.

(5) Are there any other shape invariant potentials apart from those satisfying the ansatz eq. (4.15)? We will find below that there is another ansatz based on scaling which leads to new SIPs whose potential is however only known via a Taylor series expansion.

(6) No new solutions (apart from those in Table 4.1) have been obtained so far in the case of multi-step shape invariance and when a_2 and a_1 are related by translation.

(7) What we have shown here is that shape invariance is a sufficient condition for exact solvability. But is it also a necessary condition? The answer is clearly no. Firstly, it has been shown that the solvable Natanzon potentials are in general not shape invariant. However, for the Natanzon potentials, the energy eigenvalues and wave functions are known only implicitly. Secondly there are various methods which we will discuss later of finding potentials which are strictly isospectral to the SIPs. These are not SIPs but for all of these potentials, unlike the Natanzon case, the energy eigenvalues and eigenfunctions are known in a closed form.

Before ending this subsection, we want to remark that for the SIPs (with translation) given in Table 4.1, the reflection and transmission amplitudes $R_1(k)$ and $T_1(k)$ (or phase shift $\delta_1(k)$ for the three-dimensional case) can also be calculated by operator methods. Let us first notice that since for all the cases $a_2 = a_1 + \alpha$, hence $R_1(k; a_1)$ and $T_1(k; a_1)$ are determined for all values of a_1 from eqs. (4.9) and (4.10) provided they are known in a finite strip. For example, let us consider the shape invariant superpotential

$$W = n \tanh x \ , \tag{4.49}$$

where n is a positive integer (1,2,3,...). The two partner potentials

$$\begin{aligned}
V_1(x;n) &= n^2 - n(n+1)\mathrm{sech}^2 x \ , \\
V_2(x;n) &= n^2 - n(n-1)\mathrm{sech}^2 x \ ,
\end{aligned} \tag{4.50}$$

re clearly shape invariant with

$$a_1 = n , \qquad a_2 = n - 1 . \tag{4.51}$$

On going from V_1 to V_2 to V_3 etc., we will finally reach the free particle potential which is reflectionless and for which $T = 1$. Thus we immediately conclude that the series of potentials $V_1, V_2, ...$ are all reflectionless and the transmission coefficient of the reflectionless potential $V_1(x; n)$ is given by

$$\begin{aligned}
T_1(k, n) &= \frac{(n - ik)(n - 1 - ik)...(1 - ik)}{(-n - ik)(-n + 1 - ik)...(-1 - ik)} \\
&= \frac{\Gamma(-n - ik)\Gamma(n + 1 - ik)}{\Gamma(-ik)\Gamma(1 - ik)} .
\end{aligned} \tag{4.52}$$

The scattering amplitudes for the Coulomb potential and the potential corresponding to $W = A\tanh x + B\text{sech } x$ have also been obtained in this way.

There is, however, a straightforward method for calculating the scattering amplitudes by making use of the n'th state wave functions as given in Table 4.1. In order to impose boundary conditions appropriate to the scattering problem, two modifications of the bound state wave functions have to be made: (i) instead of the parameter n labelling the number of nodes, one must use the wave number k' so that the asymptotic behavior is $\exp(ik'x)$ as $x \to \infty$ (ii) the second solution of the Schrödinger equation must be kept (it had been discarded for bound state problems since it diverged asymptotically). In this way the scattering amplitude for all the SIPs of Table 4.1 have been calculated.

4.2.2 *Solutions Involving Scaling*

From 1987 until 1993 it was believed that the only shape invariant potentials were those given in Table 4.1 and that there were no more shape invariant potentials. However, starting in 1993, a huge class of new shape invariant potentials have been discovered. It turns out that for many of these new shape invariant potentials, the parameters a_2 and a_1 are related by scaling $a_2 = qa_1, 0 < q < 1$) rather than by translation, a choice motivated by the recent interest in q-deformed Lie algebras. We shall see that many of these potentials are reflectionless and have an infinite number of bound states. So far, none of these potentials have been obtained in a closed form but are obtained only in a series form.

Let us consider an expansion of the superpotential of the from

$$W(x; a_1) = \sum_{j=0}^{\infty} g_j(x) a_1^j \ , \tag{4.53}$$

and further let

$$a_2 = q a_1, \quad 0 < q < 1 \ . \tag{4.54}$$

This is slightly misleading in that a reparameterization of the form $a_2 = q a_1$, can be recast as $a_2' = a_1' + \alpha$ merely by taking logarithms. However, since the choice of parameter is usually an integral part of constructing a SIP, it is in practice part of the ansatz. For example, we will construct below potentials by expanding in a_1, a procedure whose legitimacy and outcome are clearly dependent on our choice of parameter and hence reparameterization. We shall see that, even though the construction is non-invariant, the resulting potentials will still be invariant under redefinition of a_1. On using eqs. (4.53) and (4.54) in the shape invariance condition (4.1), writing $R(a_1)$ in the form

$$R(a_1) = \sum_{j=0}^{\infty} R_j a_1^j \ , \tag{4.55}$$

and equating powers of a_1 yields

$$2g_0'(x) = R_0 \ ; \qquad g_1'(x) + 2d_1 g_0(x) g_1(x) = r_1 d_1 \ , \tag{4.56}$$

$$g_n'(x) + 2d_n g_0(x) g_n(x) = r_n d_n - d_n \sum_{j=1}^{n-1} g_j(x) g_{n-j}(x) \ , \tag{4.57}$$

where

$$r_n \equiv R_n/(1 - q^n), \quad d_n = (1 - q^n)/(1 + q^n) \ , \qquad n = 1, 2, 3, \dots \ . \tag{4.58}$$

This set of linear differential equations is easily solvable in succession to give a general solution of eq. (4.1). Let us first consider the special case $g_0(x) = 0$, which corresponds to $R_0 = 0$. The general solution of eq. (4.57) then turns out to be

$$g_n(x) = d_n \int dx [r_n - \sum_{j=1}^{n-1} g_j(x) g_{n-j}(x)] \ , \quad n = 1, 2, \dots \tag{4.59}$$

where without loss of generality we have assumed the constants of integration to be zero. We thus see that once a set of r_n are chosen, then the shape invariance condition essentially fixes the $g_n(x)$ (and hence $W(x; a_1)$) and determines the shape invariant potential. Implicit constraints on this choice are that the resulting ground state wave function be normalizable and the spectrum be sensibly ordered which is ensured if $R(q^n a_1) > 0$.

The simplest case is $r_1 > 0$ and $r_n = 0, n \geq 2$. In this case eq. (4.59) takes a particularly simple form

$$g_n(x) = \beta_n x^{2n-1} , \qquad (4.60)$$

where

$$\beta_1 = d_1 r_1 , \qquad \beta_n = -\frac{d_n}{(2n-1)} \sum_{j=1}^{n-1} \beta_j \beta_{n-j} , \qquad (4.61)$$

and hence

$$W(x; a_1) = \sum_{j=1}^{\infty} \beta_j a_1^j x^{2j-1} = \sqrt{a_1} F(\sqrt{a_1} x) . \qquad (4.62)$$

For $a_2 = q a_1$, this gives

$$W(x; a_2) = \sqrt{q} W(\sqrt{q} x, a_1) , \qquad (4.63)$$

which corresponds to a self-similar W. It may be noted here that instead of choosing $r_n = 0, n \geq 2$, if any one r_n (say r_j) is taken to be nonzero then one again obtains self-similar potentials and in these instances the results obtained from shape invariance and self-similarity are entirely equivalent and the self-similarity condition (4.63) turns out to be a special case of the shape invariance condition.

It must be emphasized here that shape invariance is a much more general concept than self-similarity. For example, if we choose more than one r_n to be nonzero, then SIPs are obtained which are not contained within the self-similar ansatz. Consider for example, $r_n = 0, n \geq 3$. Using eq. (4.59) one can readily calculate all the $g_n(x)$, of which the first three are

$$g_1(x) = d_1 r_1 x , \qquad g_2(x) = d_2 r_2 x - \frac{1}{3} d_1^2 r_1^2 x^3 ,$$

$$g_3(x) = -\frac{2}{3} d_1 r_1 d_2 r_2 d_3 x^3 + \frac{2}{15} d_1^3 r_1^3 d_2 d_3 x^5 . \qquad (4.64)$$

Notice that in this case $W(x)$ contains only odd powers of x. This makes the potentials $V_{1,2}(x)$ symmetric in x and also guarantees unbroken SUSY. The energy eigenvalues follow immediately from eqs. (4.6) and (4.55) and are given by $(0 < q < 1)$

$$E_n^{(1)}(a_1) = \Gamma_1 \frac{(1+q)(1-q^n)}{(1-q)} + \Gamma_2 \frac{(1+q^2)(1-q^{2n})}{(1-q^2)} , n = 0, 1, 2, \dots ,$$
(4.65)

where $\Gamma_1 = d_1 r_1 a_1, \Gamma_2 = d_2 r_2 a_1^2$. The unnormalized ground state wave function is

$$\psi_0^{(1)}(x; a_1) = \exp[-\frac{x^2}{2}(\Gamma_1 + \Gamma_2) + \frac{x^4}{4}(d_2\Gamma_1^2 + 2d_3\Gamma_1\Gamma_2 + d_4\Gamma_2^2) + 0(x^6)] .$$
(4.66)

The wave functions for the excited states can be recursively calculated from the relation (4.8).

We can also calculate the transmission coefficient of this symmetric potential $(k = k')$ by using the relation (4.10) and the fact that for this SIP $a_2 = qa_1$. Repeated application of the relation (4.10) gives

$$T_1(k; a_1) = \frac{[ik - W(\infty, a_1)][ik - W(\infty, a_2)]\dots[ik - W(\infty, a_n)]}{[ik + W(\infty, a_1)][ik + W(\infty, a_2)]\dots[ik + W(\infty, a_n)]} T_1(k; a_{n+1})$$
(4.67)

where

$$W(\infty, a_j) = \sqrt{E_\infty^{(1)} - E_j^{(1)}} .$$
(4.68)

Now, as $n \to \infty, a_{n+1} = q^n a_1 \to 0(0 < q < 1)$ and, since we have taken $g_0(x) = 0$, one gets $W(x; a_{n+1}) \to 0$. This corresponds to a free particle for which the reflection coefficient $R_1(k; a_1)$ vanishes and hence the transmission coefficient of this symmetric potential is given by

$$T_1(k; a_1) = \prod_{j=1}^{\infty} \frac{[ik - W(\infty, a_j)]}{[ik + W(\infty, a_j)]} .$$
(4.69)

The above discussion keeping only $r_1, r_2 \neq 0$ can be readily generalized to an arbitrary number of nonzero r_j. The energy eigenvalues for this case are given by $(\Gamma_j \equiv d_j r_j a_1^j)$

$$E_n^{(1)}(a_1) = \sum_j \Gamma_j \frac{(1+q^j)(1-q^{jn})}{(1-q^j)}, \quad n = 0, 1, 2, \dots .$$
(4.70)

All these potentials are also symmetric and reflectionless with T_1 as given by eq. (4.69). The limits $q \to 0$ and $q \to 1$ of all these potentials are simple and quite interesting. At $q = 1$, the solution of the shape invariance condition (4.1) is the standard one dimensional harmonic oscillator with $V(x) = Rx/2$ while in the limit $q \to 0$ the solution is the Rosen-Morse superpotential corresponding to the one soliton solution given by

$$W(x) = \sqrt{R}\tanh(\sqrt{R}x) . \qquad (4.71)$$

Hence the general solution as obtained above with $0 < q < 1$ can be regarded as the multi-parameter deformation of the hyperbolic tangent function with q acting as the deformation parameter. It is also worth noting that the number of bound states increase discontinuously from just one at $q = 0$ to infinity for $q > 0$. Further, whereas for $q = 1$ the spectrum is purely discrete, for q even slightly less than one, we have the discrete as well as the continuous spectra.

Finally, let us consider the solution to the shape invariance condition (4.1) in the case when $R_0 \neq 0$. From eq. (4.56) it then follows that $g_0(x) = R_0 x/2$ rather than being zero. One can again solve the set of linear differential equations (4.56) and (4.57) in succession yielding $g_1(x), g_2(x),...$. Further, the spectrum can be immediately obtained by using eqs. (4.6) and (4.55). For example, in the case of an arbitrary number of nonzero R_j (in addition to R_0), it is given by

$$E_n^{(1)} = nR_0 + \sum_j \Gamma_j \frac{(1 + q^j)(1 - q^{jn})}{(1 - q^j)} , \qquad (4.72)$$

which is the spectrum of a q-deformed harmonic oscillator. Unlike the usual q-oscillator where the space is noncommutative but the potential is normal ($\omega^2 x^2$), in our approach the space is commutative, but the potential is deformed, giving rise to a multi-parameter deformed oscillator spectrum.

An unfortunate feature of the new SIPs obtained above is that they are not explicitly known in terms of elementary functions but only as a Taylor series about $x = 0$. Questions about series convergence naturally arise. Numerical solutions pose no serious problems. As a consistency check, Barclay et al. have checked numerically that the Schrödinger equation solved with numerically obtained potentials indeed has the analytical energy eigenvalues given above. From numerical calculations one finds that the superpotential and the potential are as shown in Figs. 4.1 and 4.2

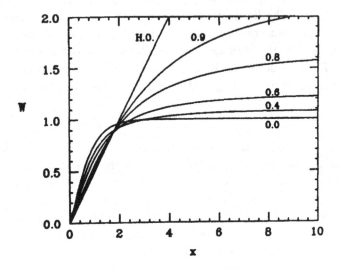

Fig. 4.1   Self-similar superpotentials $W(x)$ for various values of the deformation pa
rameter q. The curve labeled H.O. (harmonic oscillator) corresponds to the limitin
case of $q = 1$. Note that only the range $x \geq 0$ is plotted since the superpotentials ar
antisymmetric: $W(x) = -W(-x)$.

corresponding to the case when $r_1 \neq 0, r_n = 0, n \geq 2$.

A very unusual new shape invariant potential has also been obtaine(
corresponding to $r_1 = 1, r_2 = -1, r_n = 0, n \geq 3$ (with $q = 0.3$ and $a = 0.75$
which is shown in Fig. 4.3. In this case, whereas $V_1(x)$ is a double wel
potential, its shape invariant partner potential $V_2(x)$ is a single well.

It is worth pointing out that even though the potentials are not know
in a closed form in terms of elementary functions, the fact that these ar
reflectionless symmetric potentials can be used to constrain them quit
strongly. This is because, if we regard them as a solution of the K-d\
equation at time $t = 0$, then being reflectionless, it is well known that a
$t \to \pm\infty$, such solutions will break up into an infinite number of soliton
of the form $2k_i^2 \mathrm{sech}^2 k_i x$. On using the fact that the KdV solitons obey a
infinite number of conservation laws corresponding to mass, momentur
energy ..., one can immediately obtain constraints on the reflectionless SIP
obtained above.

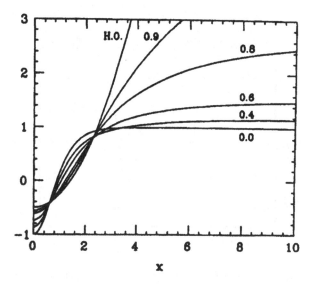

Fig. 4.2 Self-similar potentials $V_1(x)$ (symmetric about $x = 0$) corresponding to the superpotentials shown in Fig. 4.1

4.2.3 *Other Solutions*

So far we have obtained solutions where a_2 and a_1 are related either by scaling or by translation. Are there shape invariant potential where a_2 and a_1 are neither related by scaling nor by translation? It turns out that there are other possibilities for obtaining new shape invariant potentials. Some of the other possibilities are: $a_2 = qa_1^p$ with $p = 2,3,...$; $a_2 = qa_1/(1 + pa_1)$ and cyclic SIPs. Let us first consider the case when

$$a_2 = qa_1^2 , \qquad (4.73)$$

i.e. $p = 2$. Generalization to arbitrary p is straightforward. On using eqs. (4.53) and (4.55) one obtains the set of equations

$$g'_{2m}(x) + \sum_{j=0}^{2m} g_j(x)g_{2m-j}(x) = q^m \sum_{j=0}^{m} g_j(x)g_{m-j}(x) - q^m g'_m(x) + R_{2m} , \qquad (4.74)$$

$$g'_{2m+1}(x) + \sum_{j=0}^{2m+1} g_j(x)g_{2m+1-j}(x) = R_{2m+1} , \qquad (4.75)$$

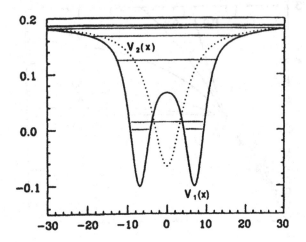

Fig. 4.3   A double well potential $V_1(x)$ (solid line) and its single well supersymmetric partner $V_2(x)$ (dotted line). Note that these two potentials are shape invariant with a scaling change of parameters. The energy levels of $V_1(x)$ are clearly marked.

which can be solved in succession and one can readily calculate all the $g_n(x)$. For example, when only R_1 and R_2 are nonzero, the first three g's are

$$g_1(x) = R_1x , \qquad g_2(x) = (R_2 - qR_1)x - \frac{1}{3}R_1^3x^3 ,$$

$$g_3(x) = \frac{2}{3}R_1(qR_1 - R_2)x^3 + \frac{2}{15}R_1^3x^5 . \qquad (4.76)$$

The corresponding spectrum turns out to be $(E_0^{(1)}(a_1) = 0)$

$$E_n^{(1)} = \frac{R_1}{q}\sum_{j=1}^{n}(a_1q)^{2^{j-1}} + \frac{R_2}{q^2}\sum_{j=1}^{n}(a_1q)^{2^j} , \; n = 1, 2, \ldots . \qquad (4.77)$$

The $q \to 0$ limit of these equations again correspond to the Rosen-Morse potential corresponding to the one soliton solution. One can also consider shape invariance in multi-steps along with this ansatz thereby obtaining deformations of the multi-soliton Rosen-Morse potential.

One can similarly consider solutions to the shape invariance condition

4.1) for the case

$$a_2 = \frac{qa_1}{1 + pa_1} \; , \qquad (4.78)$$

when $0 < q < 1$ and $pa_1 \ll 1$, so that one can expand $(1 + pa_1)^{-1}$ in powers of a_1. For example, when only R_1 and R_2 are nonzero, then one can show that the first two nonzero g_n are

$$g_1(x) = \frac{R_1 x}{(1 + q)} \; , \quad g_2(x) = (R_2 + \frac{pqR_1}{(1 + q)})x - \frac{(1 - q)}{(1 + q)^2(1 + q^2)}\frac{x^3}{3} \; , \quad (4.79)$$

and the energy eigenvalue spectrum is $(E_0^{(1)} = 0)$

$$E_n^{(1)} = R_1 \sum_{j=1}^{n} \frac{q^{j-1}a_1}{[1 + pa_1(\frac{1-q^{j-1}}{1-q})]} + R_2 \sum_{j=1}^{n} \frac{(q^{j-1}a_1)^2}{[1 + pa_1(\frac{1-q^{j-1}}{1-q})]^2} \; . \qquad (4.80)$$

Generalization to the case when several R_j are nonzero as well as shape invariance in multi-steps is straight forward.

Finally, let us consider cyclic SIPs. In this case, the SUSY partner Hamiltonians correspond to a series of SIPs which repeat after a cycle of p (p = 2,3,4,...) iterations, i.e. in this case

$$f^p(a_1) = a_1 \; . \qquad (4.81)$$

Note that here $a_2 = f(a_1), a_3 = f^2(a_1)$ etc. It has been shown that such potentials have an infinite number of periodically spaced eigenvalues. Again, in these cases the potentials are only known formally as a Taylor series except when p=2 when the potential is known in a closed form.

We would like to close this subsection with several comments.

(1) Just as we have obtained q-deformations of the reflectionless Rosen-Morse and harmonic oscillator potentials, can one also obtain deformations of the other SIPs given in Table 4.1?

(2) Have we exhausted the list of SIPs? We now have a significantly expanded list but it is clear that the possibilities are far from exhausted. In fact it appears that there are an unusually large number of shape invariant potentials, for all of which the whole spectrum can be obtained algebraically. How does one classify all these potentials? Do these potentials include all solvable potentials?

(3) For those SIPs where a_2 and a_1 are not related by translation, the spectrum has so far only been obtained algebraically. Can one directly solve the Schrödinger equation for these potentials?

(4) There is a fundamental difference between those shape invariant potentials for which a_2 and a_1 are related by translation and other choices (like $a_2 = qa_1$). In particular, whereas in the former case the potentials are explicitly known in a closed form in terms of simple functions, in the other cases they are only known formally as a Taylor series. Secondly, whereas in the latter case, all the SIPs obtained so far have infinite number of bound states and are either reflectionless or have no scattering, in the former case one has also many SIPs with nonzero reflection coefficient.

4.3   Shape Invariance and Noncentral Solvable Potentials

We have seen that using the ideas of SUSY and shape invariance, a number of potential problems can be solved algebraically. Most of these potentials are either one dimensional or are central potentials which are again essentially one dimensional but on the half line. It may be worthwhile to enquire if one can also algebraically solve some noncentral but separable potential problems. As has been shown recently, the answer to the question is yes. It turns out that the problem is algebraically solvable so long as the separated problems for each of the coordinates belong to the class of SIPs. As an illustration, let us discuss noncentral separable potentials in spherical polar coordinates.

In spherical polar coordinates (r, θ, ϕ), the Schrödinger equation is separable for a potential of the form

$$V(r, \theta, \phi) = V_1(r) + \frac{V_2(\theta)}{r^2} + \frac{V_3(\phi)}{r^2 \sin^2 \theta} , \qquad (4.82)$$

where $V_1(r), V_2(\theta)$ and $V_3(\phi)$ are arbitrary functions of their argument. The equation for the wave function $\psi(r, \theta, \phi)$ is

$$\left[-\left(\frac{\partial^2 \psi}{\partial^2 r} + \frac{2}{r}\frac{\partial \psi}{\partial r}\right) - \frac{1}{r^2}\left(\frac{\partial^2 \psi}{\partial \theta^2} + \cot\theta \frac{\partial \psi}{\partial \theta}\right) - \frac{1}{r^2 \sin^2 \theta}\frac{d^2 \psi}{\partial \phi^2} \right] = (E - V)\psi . \quad (4.83)$$

t is convenient to write $\psi(r,\theta,\phi)$ as

$$\psi(r,\theta,\phi) = \frac{R(r)}{r} \frac{H(\theta)}{(\sin\theta)^{1/2}} K(\phi) . \tag{4.84}$$

Substituting eq. (4.84) in eq. (4.83) and using the standard separation of variables procedure, one obtains the following equations for the functions $K(\phi), H(\theta)$ and $R(r)$:

$$-\frac{d^2 K}{d\phi^2} + V_3(\phi)K(\phi) = m^2 K(\phi) , \tag{4.85}$$

$$-\frac{d^2 H}{d\theta^2} + [V_2(\theta) + (m^2 - \frac{1}{4})\operatorname{cosec}^2\theta] H(\theta) = l^2 H(\theta) , \tag{4.86}$$

$$-\frac{d^2 R}{dr^2} + [V_1(r) + \frac{(l^2 - \frac{1}{4})}{r^2}] R(r) = E R(r) , \tag{4.87}$$

where m^2 and l^2 are separation constants.

The three Schrödinger equations given by (4.85), (4.86) and (4.87) may be solved algebraically by choosing appropriate SIPs for $V_3(\phi), V_2(\theta)$ and $V_1(r)$.

Generalization of this technique to noncentral but separable potentials n other orthogonal curvilinear coordinate systems as well as in other dimensions is quite straightforward.

References

(1) L. Gendenshtein, *Derivation of Exact Spectra of the Schrödinger Equation by Means of Supersymmetry*, JETP Lett. **38** (1983) 356-358.

(2) E. Schrödinger, *A Method of Determining Quantum-Mechanical Eigenvalues and Eigenfunctions*, Proc. Roy. Irish Acad. **A46** (1940) 9-16.

(3) L. Infeld and T.E. Hull, *The Factorization Method*, Rev. Mod. Phys. **23** (1951) 21-68.

(4) G.A. Natanzon, *Study of the One Dimensional Schrödinger Equation Generated from the Hypergeometric Equation*, Vestnik Leningrad Univ. **10** (1971) 22-28; *General Properties of Potentials for which*

the *Schrödinger Equation Can be Solved by Means of Hypergeometric Functions*, Teoret. Mat. Fiz. **38** (1979) 219-229.

(5) G. A. Natanzon, *Construction of the Jost Function and of the S-Matrix for a General Potential*, Sov. J. Phys. **21** (1978) 855-859.

(6) J.N. Ginocchio, *A Class of Exactly Solvable Potentials I,II*, Ann. Phys. **152** (1984) 203-219; ibid. **159** (1985) 467-480.

(7) R. Dutt, A. Khare and U.P. Sukhatme, *Supersymmetry, Shape Invariance and Exactly Solvable Potentials*, Am. J. Phys. **56** (1988) 163-168.

(8) F. Cooper, J.N. Ginocchio and A. Khare, *Relationship Between Supersymmetry and Solvable Potentials*, Phys. Rev. **D36** (1987) 2458-2473.

(9) D. Barclay, R. Dutt, A. Gangopadhyaya, A. Khare, A. Pagnamenta and U.P. Sukhatme, *New Exactly Solvable Hamiltonians: Shape Invariance and Self-Similarity*, Phys. Rev. **A48** (1993) 2786-2797.

(10) U.P. Sukhatme, C. Rasinariu and A. Khare, *Cyclic Shape Invariant Potentials*, Phys. Lett. **A234** (1997) 401-409.

(11) A. Khare and R.K. Bhaduri, *Supersymmetry, Shape Invariance and Exactly Solvable Noncentral Potentials*, Am. J. Phys. **62** (1994) 1008-1014.

(12) A. Khare and R.K. Bhaduri, *Some Algebraically Solvable Three-Body Problems in One Dimension*, J. Phys. **A27** (1994) 2213-2223.

(13) J.F. Cariñena and A. Ramos, *Shape Invariant Potentials Depending on n Parameters Transformed by Translation*, J. Phys. **A33** (2000) 3467-3481.

(14) R. Dutt, A. Gangopadhyaya, and U. Sukhatme, *Noncentral Potentials and Spherical Harmonics Using Supersymmetry and Shape Invariance*, Am. J. Phys. **65** (1977) 400-403.

(15) B. Gönül and I. Zorba, *Supersymmetric Solutions of Non-Central Potentials*, Phys. Lett. **A269** (2000) 83-88.

Problems

1. Consider the ground state wave function $\psi_0(r) = Ar^3 e^{-r^2/r_0^2}$. (i) Taking $r_0 = 1$, plot $\psi_0(r)$ versus r. (ii) Compute and graph the superpotential $V(r)$ and the partner potentials $V_1(r)$ and $V_2(r)$. (iii) Show that $V_1(r)$ is shape invariant and find its eigenvalues. (iv) Compute and plot the first and second excited states of $V_1(r)$. (v) Compute and plot the ground and first excited states of $V_2(r)$.

2. Make a list of all known symmetric shape invariant potentials in which the change of parameters is a translation. Give the corresponding superpotentials, and the energies of the three lowest eigenstates. Take units with $\hbar = 2m = 1$.

3. The Hulthén potential

$$V(r) = -V_0 \frac{e^{-r/a}}{1 - e^{-r/a}}$$

is widely used in atomic physics. (i) Plot the potential as a function of the dimensionless parameter r/a. (ii) Show that the Hulthén potential can be re-cast in the form of the Eckart potential given in the list of shape invariant potentials. (iii) What are the energy eigenvalues of the Hulthén potential?

Chapter 5

Charged Particles in External Fields and Supersymmetry

5.1 Spinless Particles

To obtain the Schrödinger equation for a particle of charge q in external electric and magnetic fields we must first find the Lagrangian and Hamiltonian from which the equation of motion with a Lorentz force law

$$m\frac{d\vec{v}}{dt} = q(\vec{E} + \frac{1}{c}\vec{v} \times \vec{B}) , \tag{5.1}$$

can be obtained. Following standard textbook methods such as that found in Goldstein, the appropriate Lagrangian is

$$L = \frac{1}{2m}(m\vec{v} + \frac{q}{c}\vec{A})^2 - \frac{q^2}{2mc^2}\vec{A} \cdot \vec{A} - q\phi , \tag{5.2}$$

where $\vec{A}$ and ϕ are the vector and scalar gauge potentials. The momentum conjugate to $\vec{x}$ is

$$\vec{p} = \frac{\partial L}{\partial \vec{v}} = m\vec{v} + \frac{q}{c}\vec{A} . \tag{5.3}$$

Using the standard Legendre transformation from the variables $\{x, \dot{x}\}$ to $\{x, p\}$ we obtain

$$H = \vec{v} \cdot \vec{p} - L = \frac{1}{2m}(\vec{p} - \frac{q}{c}\vec{A})^2 + q\phi . \tag{5.4}$$

Using the correspondence principle

$$\vec{p} \to \frac{\hbar}{i}\vec{\nabla} , \quad H \to i\hbar\frac{\partial}{\partial t} ,$$

61

we obtain the Schrödinger equation:

$$i\hbar \frac{\partial \psi}{\partial t} = H\psi \ , \tag{5.5}$$

where

$$H = \frac{1}{2m}(\frac{\hbar}{i}\vec{\nabla} - \frac{q}{c}\vec{A})^2 + q\phi \ .$$

In a uniform magnetic field $\vec{B}$ the vector potential is given by

$$\vec{A} = \frac{1}{2}\vec{B} \times \vec{r} \ , \tag{5.6}$$

and the time independent Schrödinger equation becomes

$$\left[-\frac{\hbar^2}{2m}\nabla^2 - \frac{q}{2mc}\vec{B}\cdot\vec{L} + \frac{q^2}{8mc^2}(\vec{B}\times\vec{r})^2 + q\phi\right]\psi = E\psi \ . \tag{5.7}$$

Here

$$\vec{L} = \vec{r} \times \frac{\hbar}{i}\vec{\nabla} \ .$$

The first term in the potential energy is the interaction of a magnetic field with a magnetic dipole of form

$$\vec{\mu} = \frac{q}{2mc}\vec{L} \ . \tag{5.8}$$

The second term is a term quadratic in $\vec{r}$ and leads to the second order Zeeman effect for atoms in external magnetic fields. This Hamiltonian does not give the correct spectrum for atoms in an external magnetic field since it ignores the intrinsic spin of the electron.

5.2   Non-relativistic Electrons and the Pauli Equation

To describe electrons non-relativistically, Pauli introduced the concept of intrinsic spin and extended the wave function to include a spin quantum number m_s. Wave functions are now defined on a product space of orbital angular momentum and intrinsic angular momentum or spin. The representations of angular momentum for spin $1/2$ are best described in terms of the Pauli matrices $\vec{\sigma} = 2\vec{s}$, with components:

$$\sigma_1 = \begin{pmatrix} 0 & 1 \\ 1 & 0 \end{pmatrix} \ , \quad \sigma_2 = \begin{pmatrix} 0 & -i \\ i & 0 \end{pmatrix} \ , \quad \sigma_3 = \begin{pmatrix} 1 & 0 \\ 0 & -1 \end{pmatrix} \ . \tag{5.9}$$

We have

$$\vec{s} \cdot \vec{s} = \begin{pmatrix} 3/4 & 0 \\ 0 & 3/4 \end{pmatrix} = s(s+1)I , \tag{5.10}$$

and I is the unit matrix in 2 dimensions

$$I = \begin{pmatrix} 1 & 0 \\ 0 & 1 \end{pmatrix} .$$

The spin degrees of freedom are described by the two eigenstates of σ_3

$$\chi(m_s = \pm\frac{1}{2}) = \chi^{\pm} ; \quad \sigma_3 \chi^{\pm} = \pm\chi^{\pm} . \tag{5.11}$$

If one is in a central field the wave functions are product wave functions of the type

$$\Psi(l m_l m_s) = R_l(r) \, Y_l^{m_l}(\theta, \phi) \, \chi(m_s) , \tag{5.12}$$

where the $Y_l^m(\theta, \phi)$ are the standard spherical harmonics. For discussing supersymmetry it is useful to also introduce the raising and lowering operators:

$$\sigma_+ = \frac{1}{2}(\sigma_1 + i\sigma_2) = \begin{pmatrix} 0 & 1 \\ 0 & 0 \end{pmatrix} , \quad \sigma_- = \frac{1}{2}(\sigma_1 - i\sigma_2) = \begin{pmatrix} 0 & 0 \\ 0 & 1 \end{pmatrix} . \tag{5.13}$$

If one now has an electron in a purely magnetic field there is an additional interaction of the form (for the electron we set $q = -e$ with $e > 0$)

$$H_I = \frac{g}{2} \frac{e\hbar}{2mc} \vec{B} \cdot \vec{\sigma} , \tag{5.14}$$

now acting on a two component wave function. Note that the electron has an intrinsic magnetic moment

$$|\,\mu_s\,| = \frac{e\hbar}{2mc} = 0.9273 \times 10^{-20} \text{erg/gauss} = 1 \text{ Bohr magneton} . \tag{5.15}$$

A gyromagnetic ratio of $g = 2$ results naturally from an $N = 1$ supersymmetry of the Pauli equation and also from the Dirac equation. The Pauli Hamiltonian is explicitly (at $g = 2$)

$$H = \frac{1}{2m} \left(\vec{p} + \frac{e}{c}\vec{A} \right)^2 + \frac{e\hbar}{2mc} \vec{B} \cdot \vec{\sigma} . \tag{5.16}$$

This equation already has an $N = 1$ supersymmetry if we introduce a single self-adjoint ($N = 1$) supercharge:

$$Q_1 = \frac{1}{\sqrt{4m}} \left(\vec{p} + \frac{e}{c}\vec{A} \right) \cdot \vec{\sigma} . \tag{5.17}$$

This is because the Pauli Hamiltonian eq. (5.16) can be written as

$$H = 2Q_1^2 = \{Q_1, Q_1\} , \tag{5.18}$$

and obviously

$$[H, Q_1] = 0 . \tag{5.19}$$

When the magnetic field is perpendicular to the motion of the electron, there is instead an $N = 2$ supersymmetry which then relates this problem to a 1-D SUSY quantum mechanics problem. There are several ways of introducing the $N = 2$ supersymmetry. The most symmetric form having $N = 2$ supersymmetry is to introduce the complex supercharge

$$Q = \sqrt{2/m} \left[(p_x + \frac{e}{c}A_x) - i(p_y + \frac{e}{c}A_y) \right] \sigma_+ \equiv A\sigma_+ , \tag{5.20}$$

which obeys the superalgebra

$$Q^2 = 0 , \quad \{Q, Q^\dagger\} = H_P , \tag{5.21}$$

where H_P denotes the Pauli Hamiltonian for the special case when the motion of the electron is in a plane perpendicular to the magnetic field i.e.

$$H_P = \frac{1}{2m} \sum_{i=1}^{2} (p_i + \frac{e}{c}A_i)^2 + \frac{e\hbar}{2mc}B_3\sigma_3 . \tag{5.22}$$

The supersymmetry then guarantees that all the positive energy eigenvalues of H_P are spin degenerate. These degenerate eigenstates are connected by the operators Q and $Q^\dagger$. The Hamiltonians acting on the two subspaces of spin up and spin down are $AA^\dagger$ and $A^\dagger A$ respectively.

Another way of writing this superalgebra is to write

$$
\begin{aligned}
Q \quad &= \frac{i}{\sqrt{2}}(Q^1 + iQ^2) , \\
Q^1 \quad &= \frac{1}{\sqrt{2}} \left[-(p_y + \frac{e}{c}A_y)\sigma_1 + (p_x + \frac{e}{c}A_x)\sigma_2 \right] , \\
Q^2 \quad &= \frac{1}{\sqrt{2}} \left[(p_x + \frac{e}{c}A_x)\sigma_1 + (p_y + \frac{e}{c}A_y)\sigma_2 \right] .
\end{aligned}
\tag{5.23}
$$

For this choice of Q^i

$$\{Q^a, Q^b\} = H_P \delta^{ab} , \quad [H_P, Q^a] = 0 , \quad a, b = 1, 2 . \qquad (5.24)$$

To find exact solutions to the Pauli equation, it is useful to exploit gauge invariance to figure out simple choices of the vector potential for the external magnetic field which will lead to an exactly solvable potential problem. Since $B_z(x, y) = \frac{\partial A_y}{\partial x} - \frac{\partial A_x}{\partial y}$, different choices of $\vec{A}$ differing by a total gradient lead to the same magnetic field. By being clever we can reduce the problem trivially to a one dimensional one (without using polar coordinates) and then use our previous results on shape invariant potentials.

First to simplify things we will choose our dimensional units such that ($\hbar = 2m = e = c = 1$). Then the Pauli Hamiltonian for the motion of a charged particle in a plane (here $x - y$) in an external magnetic field in the direction perpendicular to that plane (here z direction) becomes

$$H_P = (p_x + A_x)^2 + (p_y + A_y)^2 + B_z \sigma_3 . \qquad (5.25)$$

The Hamiltonian eq. (5.25) has an additional $O(2) \times O(2)$ symmetry coming from σ_3 and an $O(2)$ rotation in the $A^1 - A^2$ plane, where $A^1 = p_x + A_x, A^2 = p_y + A_y$. One can analyze the solvable potentials most simply in an asymmetric gauge where we choose one of A_x, A_y to be zero, and the other to be a function of the opposite variable. i.e.

$$A_y(x, y) = 0 , \quad A_x(x, y) = W(y) , \qquad (5.26)$$

so that

$$B_z = -\frac{dW(y)}{dy} .$$

This will lead to three different possible solvable potential problems, whereas a symmetric choice of gauge leads naturally only to the case of the uniform field which is essentially a harmonic oscillator as we shall show below. In the asymmetric gauge the Pauli Hamiltonian takes the form

$$H_P = (p_x + W(y))^2 + p_y^2 - W'(y)\sigma_3 . \qquad (5.27)$$

Since this H_P does not depend on x, hence the eigenfunction $\tilde{\psi}$ can be factorized as

$$\tilde{\psi}(x, y) = e^{ikx}\psi(y) , \qquad (5.28)$$

where k is the eigenvalue of the operator p_x ($-\infty \le k \le \infty$). The Schrödinger equation for $\psi(y)$ then takes the form

$$[-\frac{d^2}{dy^2} + (W(y) + k)^2 - W'(y)\sigma]\psi(y) = E\psi(y) , \qquad (5.29)$$

where $\sigma(= \pm 1)$ is the eigenvalue of the operator σ_3. Thus we have reduced the problem to that of SUSY in one dimension with superpotential $W(y)+k$ where $W(y)$ must be independent of k. This constraint on $W(y)$ strongly restricts the allowed forms of shape invariant $W(y)$ for which the spectrum can be written down algebraically. In particular, from our previous discussion of the one-dimensional Schrödinger equation we find that the only allowed forms are

(i) $W(y) = \omega_c y + c_1$
(ii) $W(y) = a \tanh y + c_1$
(iii) $W(y) = a \tan y + c_1, \quad -\frac{\pi}{2} \le y \le \frac{\pi}{2}$
(iv) $W(y) = c_1 - c_2 \exp(-y)$

for which $W(y)$ can be written in terms of simple functions and for which the spectrum can be written down algebraically. In particular, for case (i),

$$W(y) = \omega_c y + c_1 , \qquad (5.30)$$

which corresponds to the case of uniform magnetic field. For this case, the energy eigenvalues are known as Landau levels, and are given by

$$E_n = (2n + 1 + \sigma)\omega_c , \quad n = 0, 1, 2... . \qquad (5.31)$$

Note that the ground state and all excited states are infinite-fold degenerate since E_n does not depend on k which assumes a continuous sequence of value ($-\infty \le k \le \infty$).

The magnetic field corresponding to the other choices of W are

(ii) $B = -a \operatorname{sech}^2 y$
(iii) $B = -a \sec^2 y \quad (-\frac{\pi}{2} \le y \le \frac{\pi}{2})$
(iv) $B = c_2 \exp(-y)$

and as mentioned above, all these problems can be solved algebraically.

Let us now consider the same problem in the symmetric gauge. We choose

$$A_x = \omega_c y f(\rho) , \quad A_y = -\omega_c x f(\rho) , \qquad (5.32)$$

where $\rho^2 = x^2 + y^2$ and ω_c is a constant. The corresponding magnetic field B_z is then given by

$$B_z(x, y) \equiv \partial_x A_y - \partial_y A_x = -2\omega_c f(\rho) - \omega_c \rho f'(\rho) . \tag{5.33}$$

In this case the Pauli Hamiltonian can be shown to take the form

$$H = -\left(\frac{d^2}{dx^2} + \frac{d^2}{dy^2}\right) + \omega_c^2 \rho^2 f^2 - 2\omega_c f L_z - (2\omega_c f + \omega_c \rho f')\sigma_3 , \tag{5.34}$$

where L_z is the z-component of the orbital angular momentum operator. Clearly the corresponding Schrödinger problem can be solved in the cylindrical coordinates ρ, ϕ. In this case, the eigenfunction $\psi(\rho, \phi)$ can be factorized as

$$\psi(\rho, \phi) = R(\rho)e^{im\phi}/\sqrt{\rho} , \tag{5.35}$$

where $m = 0, \pm 1, \pm 2, \ldots$ is the eigenvalue of L_z. In this case the Schrödinger equation for $R(\rho)$ takes the form

$$\left[-\frac{d^2}{d\rho^2} + \omega_c^2 \rho^2 f^2 - 2\omega_c f m - (2\omega_c f + \omega_c \rho f'(\rho))\sigma + \frac{m^2 - 1/4}{\rho^2}\right] R(\rho) = ER(\rho) , \tag{5.36}$$

where $\sigma(= \pm 1)$ is the eigenvalue of the operator σ_3. There is one shape invariant potential $(f(\rho) = 1)$ for which the spectrum can be written down algebraically. This case again corresponds to the famous Landau level problem i.e. it corresponds to the motion of a charged particle in the $x - y$ plane and subjected to a uniform magnetic field (in the symmetric gauge) in the z-direction. The energy eigenvalues are

$$E_n = 2(n + m + \mid m \mid)\omega_c , \quad n = 0, 1, 2\ldots , \tag{5.37}$$

so that all the states are again infinite-fold degenerate. It is worth noting that nonuniform magnetic fields can also give this equi-spaced spectrum. However, they do so only for one particular value of m while for other values of m, the spectrum is in general not equi-spaced.

The fact that in this example there are infinite number of degenerate ground states with zero energy can be understood from the Aharonov-Casher theorem which states that if the total flux defined by

$$\Phi = \int B_z dx dy = n + \epsilon \ (0 \leq \epsilon < 1) ,$$

then there are precisely $n - 1$ zero energy states. Note that in our case Φ is infinite.

5.3   Relativistic Electrons and the Dirac Equation

The free relativistic electron obeys a 4-component wave equation known as the Dirac equation. In units where ($\hbar = c = 1$) the equation can be written

$$(i\gamma^\mu \partial_\mu - m)\,\psi = 0 \ . \tag{5.38}$$

The γ^μ are 4×4 matrices, known as the Dirac gamma matrices and they obey the anti-commutation relations:

$$\{\gamma^\mu, \gamma^\nu\} = \gamma^\mu\gamma^\nu + \gamma^\nu\gamma^\mu = 2g^{\mu\nu}\mathbf{I} \ . \tag{5.39}$$

The relativistic metric we use is

$$g^{\mu\nu} = \text{diagonal}\,(1, -1, -1, -1) \ . \tag{5.40}$$

The objects

$$\frac{\sigma_{\mu\nu}}{2} = \frac{i}{4}[\gamma_\mu, \gamma_\nu] \ ,$$

transform as generators of the inhomogeneous Lorentz group and γ^μ transforms as a vector under this Lorentz group. To make a scalar out of the fields ψ one introduces the quantity

$$\bar{\psi} \equiv \psi^\dagger \gamma^0 \ , \tag{5.41}$$

and then one can show that $\bar{\psi}\psi$ transforms like a scalar under Lorentz transformations. The Lagrangian density for the Dirac equation is then written as

$$\mathcal{L} = \bar{\psi}(x)\,(i\gamma^\mu \partial_\mu - m)\,\psi(x) \ . \tag{5.42}$$

The standard representation of the gamma matrices is

$$\gamma^0 = \begin{pmatrix} I & 0 \\ 0 & -I \end{pmatrix}; \quad \gamma^i = -\gamma_i = \begin{pmatrix} 0 & \sigma_i \\ -\sigma_i & 0 \end{pmatrix} \ , \tag{5.43}$$

where I is the unit matrix, σ_i are the Pauli matrices and we have used a condensed notation that each entry is a 2×2 matrix. Another non-covariant way that the Dirac equation is written is

$$i\frac{\partial \psi}{\partial t} = \left(\frac{1}{i}\alpha \cdot \nabla + \beta m\right)\psi = H\psi .$$ (5.44)

Here $\beta = \gamma^0$ and

$$\alpha_i = \beta\gamma^i = \begin{pmatrix} 0 & \sigma_i \\ \sigma_i & 0 \end{pmatrix} .$$ (5.45)

One also has

$$\alpha_i^2 = \beta^2 = I ; \quad \{\alpha_i, \beta\} = 0 .$$ (5.46)

Therefore the Hamiltonian squared takes on the simple form:

$$H^2 = -\vec{\nabla}^2 + m^2 .$$ (5.47)

We seek plane wave solution of the free Dirac equation of the form:

$$\psi^{(+)}(\mathbf{x}, \mathbf{t}) = e^{i(\mathbf{p}\cdot\mathbf{x} - Et)}u(p, s), \quad s = \pm 1 \text{ positive energy} ,$$
$$\psi^{(-)}(\mathbf{x}, \mathbf{t}) = e^{-i(\mathbf{p}\cdot\mathbf{x} - Et)}v(p, s), \quad s = \pm 1 \text{ negative energy},$$

with the condition that E is positive. The Dirac equation implies

$$(\gamma^\mu p_\mu - m)\,u(p, s) = 0 ,$$
$$(\gamma^\mu p_\mu + m)\,v(p, s) = 0 .$$ (5.48)

In the rest frame of the particle where $\vec{p} = 0$ we obtain

$$(\gamma^0 - 1)u(m, 0) = 0; \quad (\gamma^0 + 1)v(m, 0) = 0 .$$ (5.49)

There are two linearly independent u solutions and two v's which lead to the four linearly independent solutions in the rest frame:

$$\psi_1 = e^{-imt}u(m, +) = e^{-imt}\begin{pmatrix} 1 \\ 0 \\ 0 \\ 0 \end{pmatrix} ; \psi_2 = e^{-imt}u(m, -) = e^{-imt}\begin{pmatrix} 0 \\ 1 \\ 0 \\ 0 \end{pmatrix} ;$$

$$\psi_3 = e^{imt}v(m, +) = e^{imt}\begin{pmatrix} 0 \\ 0 \\ 1 \\ 0 \end{pmatrix} ; \quad \psi_4 = e^{imt}v(m, -) = e^{imt}\begin{pmatrix} 0 \\ 0 \\ 0 \\ 1 \end{pmatrix} ,$$

(5.50)

the last two corresponding to negative energy solutions.

To obtain $u(p, s)$ at nonzero three momentum we make use of the fact that

$$(\gamma^\mu p_\mu - m)(\gamma^\mu p_\mu + m) = 0.$$

Therefore, if we define

$$u(p, s) \quad = \frac{1}{\sqrt{2m(E + m)}}(\gamma^\mu p_\mu + m)u(m, s),$$

$$v(p, s) \quad = \frac{1}{\sqrt{2m(E + m)}}(-\gamma^\mu p_\mu + m)v(m, s),$$

$$(5.51)$$

(where now in $u(p, s)$, p stands for the 4-vector $(E, \vec{p})$), then eq. (5.48) is automatically satisfied.

5.4   SUSY and the Dirac Equation

We are interested in the case of electrons in the presence of external fields, both scalar as well as electromagnetic. For many external field problems, the Dirac Hamiltonian can be put in the form (see for example the textbook of Thaller)

$$H_D = \begin{pmatrix} M_+ & Q^\dagger \\ Q & -M_- \end{pmatrix}, \tag{5.52}$$

with the following relationships being valid:

$$Q^\dagger M_- = M_+ Q^\dagger ; \quad QM_+ = M_- Q. \tag{5.53}$$

When those relationships hold, the square of the Dirac Hamiltonian becomes diagonal and of the form:

$$H_D^2 = \begin{pmatrix} Q^\dagger Q + M_+^2 & 0 \\ 0 & QQ^\dagger + M_-^2 \end{pmatrix}. \tag{5.54}$$

The Dirac Hamiltonian itself can then be diagonalized and put in the form:

$$H_D = \begin{pmatrix} \sqrt{Q^\dagger Q + M_+^2} & 0 \\ 0 & -\sqrt{QQ^\dagger + M_-^2} \end{pmatrix}, \tag{5.55}$$

showing that the positive and negative energy solutions decouple. Since the operators $QQ^\dagger$ and $Q^\dagger Q$ are isospectral, the positive and negative eigenvalues of H are closely related. Furthermore, if $Q^\dagger Q$ can be related to the Hamiltonian of a solvable SUSY quantum mechanics problem then we will be able to solve the Dirac equation exactly.

We will consider two problems below. First we will consider the Dirac equation in 1+1 dimensions with Lorentz scalar potential $\phi(x)$. Then we will consider the Dirac equation in an external electromagnetic field. For the scalar problem, the interaction Hamiltonian is obtained by replacing

$$m \to m + \phi(x) \equiv \Phi(x) \; . \tag{5.56}$$

The covariant Dirac equation becomes:

$$i\gamma^\mu \partial_\mu \psi(x,t) - \Phi(x)\psi(x,t) = 0 \; , \tag{5.57}$$

which in the non-covariant form corresponds to

$$i\frac{\partial \psi(x,t)}{\partial t} = H\psi(x,t) \; , \tag{5.58}$$

with

$$H = \alpha \cdot p + \beta\Phi(x) \; .$$

Choosing a "supersymmetric" representation of the α and β matrices,

$$\alpha_i = \begin{pmatrix} 0 & \sigma_i \\ \sigma_i & 0 \end{pmatrix} ; \beta = \begin{pmatrix} 0 & -i \\ i & 0 \end{pmatrix} \; , \tag{5.59}$$

this Hamiltonian can be put in the standard form (5.52) with

$$Q = p \cdot \sigma + i\Phi(x) \; ; \; M_+ = M_- = 0 \; . \tag{5.60}$$

The spectrum of the Hamiltonian can now be obtained from the spectrum of the operators:

$$\begin{aligned} Q^\dagger Q &= -\nabla^2 + \Phi^2 + \sigma \cdot \nabla\Phi \; , \\ QQ^\dagger &= -\nabla^2 + \Phi^2 - \sigma \cdot \nabla\Phi \; . \end{aligned} \tag{5.61}$$

If we are in 1+1 dimension, or have a potential which is only a function of one variable then this problem gets reduced to understanding solvable problems in 1-D quantum mechanics.

For an electron in an external electromagnetic field the Dirac equation is obtained by replacing the ordinary derivative by the gauge covariant derivative:

$$\partial_\mu \to \partial_\mu + iA_\mu .$$

Note that electron charge is $-e$ and we are using units where $e = \hbar = c = 1$. This gives an interaction term in the Hamiltonian:

$$H_I = \bar{\psi}\gamma^\mu A_\mu \psi . \qquad (5.62)$$

In non-covariant form the Dirac Hamiltonian for a particle in a pure magnetic field $(A_0 = 0, \vec{A} = \vec{A}(\vec{r}))$ can be written

$$H_D = \vec{\alpha} \cdot (\vec{p} + \vec{A}) + \beta m . \qquad (5.63)$$

Using the standard representation of the matrices α, β as given by eqs. (5.45) and (5.43) we can cast this Dirac Hamiltonian in the form (5.52) with the identification

$$Q = Q^\dagger = \vec{\sigma} \cdot (\vec{p} + \vec{A}) ; \quad M_\pm = m . \qquad (5.64)$$

This charge Q (apart from a factor $1/\sqrt{4m}$) is exactly the supercharge for the Pauli Hamiltonian eq. (5.17). The square of the Hamiltonian has the form:

$$H_D^2 = \begin{pmatrix} Q^\dagger Q + m^2 & 0 \\ 0 & QQ^\dagger + m^2 \end{pmatrix} , \qquad (5.65)$$

which (apart from a rescaling and a shift by the rest mass energy m) has on the diagonal two copies of the Pauli Hamiltonian. Thus when we solve the Pauli Hamiltonian in an external magnetic field we also determine a solution of the corresponding Dirac problem.

5.5   Dirac Equation with a Lorentz Scalar Potential in $1 + 1$ Dimensions

The Dirac equation in 1+1 dimension in the presence of a scalar potential is interesting because it has been used as a model for polymers such as polyacetylene. Purely scalar field theories in 1+1 dimension with quartic self interactions have finite energy kink solutions such as those found in the Korteweg-de Vries equation discussed elsewhere in this book.

The Dirac Lagrangian in 1+1 dimensions with a Lorentz scalar potential $\phi(x)$ is given by

$$\mathcal{L} = i\bar{\psi}\gamma^{\mu}\partial_{\mu}\psi - \bar{\psi}\psi\phi \ . \tag{5.66}$$

Here, the scalar potential $\phi(x)$ can be looked upon as the static, finite energy, kink solution corresponding to the scalar field Lagrangian

$$\mathcal{L}_{\phi} = \frac{1}{2}\partial_{\mu}\phi(x)\partial^{\mu}\phi(x) - V(\phi) \ . \tag{5.67}$$

The Dirac equation following from eq. (5.66) is

$$i\gamma^{\mu}\partial_{\mu}\psi(x,t) - \phi(x)\psi(x,t) = 0 \ . \tag{5.68}$$

First let us choose a two dimensional representation of the γ matrices to directly cast the problem in 1-D SUSY form. Then we will use our general formalism above to obtain solutions for this case. Let

$$\psi(x,t) = \exp(-i\omega t)\psi(x) \ , \tag{5.69}$$

so that the Dirac equation reduces to

$$\gamma^{0}\omega\psi(x) + i\gamma^{1}\frac{d\psi(x)}{dx} - \phi(x)\psi(x) = 0 \ . \tag{5.70}$$

We choose

$$\gamma^{0} = \sigma_{1} = \begin{pmatrix} 0 & 1 \\ 1 & 0 \end{pmatrix}, \ \gamma^{1} = i\sigma_{3} = \begin{pmatrix} i & 0 \\ 0 & -i \end{pmatrix}, \ \psi(x) = \begin{pmatrix} \psi_{1}(x) \\ \psi_{2}(x) \end{pmatrix} , \tag{5.71}$$

so that we have the coupled equations

$$\begin{aligned} A\psi_{1}(x) &= \omega\psi_{2}(x) \ , \\ A^{\dagger}\psi_{2}(x) &= \omega\psi_{1}(x) \ , \end{aligned} \tag{5.72}$$

where

$$A = \frac{d}{dx} + \phi(x) \ , \ A^{\dagger} = -\frac{d}{dx} + \phi(x) \ . \tag{5.73}$$

We can now easily decouple these equations. We get

$$A^{\dagger}A\psi_{1} = \omega^{2}\psi_{1} \ , \ AA^{\dagger}\psi_{2} = \omega^{2}\psi_{2} \ . \tag{5.74}$$

On comparing with the formalism of SUSY QM, we see that there is a supersymmetry in the problem and $\phi(x)$ is just the superpotential of the Schrödinger formalism. Further ψ_{1} and ψ_{2} are the eigenfunctions of the Hamiltonians $H_1 \equiv A^+A$ and $H_2 \equiv AA^\dagger$ respectively with the corresponding potentials being $V_{1,2}(x) \equiv \phi^2(x) \mp \phi'(x)$. The spectrum of the two Hamiltonians is thus degenerate except that $H_1(H_2)$ has an extra state at zero energy so long as $\phi(x \to \pm\infty)$ have opposite signs and $\phi(x \to +\infty) > 0(< 0)$.

This result could also have been obtained by specializing eq. (5.61) to one dimension and choosing $\Phi = \phi(x)$ so that the Hamiltonian squared is immediately:

$$H^2 = -\frac{d^2}{dx^2} + \phi^2(x) + \sigma_3 \frac{d\phi(x)}{dx} \ . \tag{5.75}$$

Using the results previously obtained for 1-D SUSY QM, we then conclude that for every SIP, there exists an analytically solvable Dirac problem with the corresponding scalar potential $\phi(x)$ being the superpotential of the Schrödinger problem. In particular, using the reflectionless superpotential given by

$$W(x) = n \tanh x \ , \tag{5.76}$$

one can immediately construct perfectly transparent Dirac potentials with n bound states. Further, using the results for the SIP with scaling ansatz $(a_2 = qa_1)$ one can also construct perfectly transparent Dirac potentials with an infinite number of bound states.

We can also solve the scalar Dirac Hamiltonian in higher dimensions as long as Φ depends on only one coordinate. This has been used as a model of spatially dependent valence and conduction band edges of semiconductors near the Γ and L points in the Brillouin zone. One assumes that one can write

$$\psi(x, y, z) = e^{ik_x x + ik_y y} \ \psi(z) \ , \tag{5.77}$$

so that on these wave functions (assuming $\Phi = \phi(z)$) one has that eq. (5.61) becomes

$$Q^\dagger Q = k_x^2 + k_y^2 - \frac{d^2}{dz^2} + \phi^2 + \sigma_3 \frac{d\phi}{dz} \ ,$$

$$QQ^\dagger = k_x^2 + k_y^2 - \frac{d^2}{dz^2} + \phi^2 - \sigma_3 \frac{d\phi}{dz} \ . \tag{5.78}$$

Thus the eigenvalues for the quantum mechanics problem are related directly to

$$E_n = \omega_n^2 - (k_x^2 + k_y^2) \ . \tag{5.79}$$

This problem is discussed in more detail in the book of Junker.

5.6 Supersymmetry and the Dirac Particle in a Coulomb Field

The Dirac equation for a charged particle in an electromagnetic field is given by $(e = \hbar = c = 1)$

$$[i\gamma^\mu(\partial_\mu + iA_\mu) - m]\psi = 0 \ . \tag{5.80}$$

For a central field i.e. $\vec{A} = 0$ and $A_0(\vec{x}, t) = V(r)$, the non-covariant form of the equation is

$$i\frac{\partial\psi}{\partial t} = H\psi = (\vec{\alpha}\cdot\vec{p} + \beta m + V)\psi \ . \tag{5.81}$$

For central fields, this Dirac equation can be separated in spherical coordinates. That is we can find simultaneous eigenstates of the total angular momentum $\vec{J}^2$ as well as J_z and H. If we construct the 4- component wave function ψ in terms of the two component spinors ϕ and χ

$$\psi = \begin{bmatrix} \phi \\ \psi \end{bmatrix} \ , \tag{5.82}$$

then the two component angular eigenfunctions are:

$$\phi_{j,m}^{(\pm)} = \begin{bmatrix} \sqrt{\frac{l+1/2\pm m}{2l+1}} Y_l^{m-1/2} \\ \pm\sqrt{\frac{l+1/2\mp m}{2l+1}} Y_l^{m+1/2} \end{bmatrix} \ , \tag{5.83}$$

corresponding to whether $j = l \pm 1/2$. These solutions satisfy the eigenvalue equation:

$$\vec{J}^2\phi_{j,m}^{(\pm)} = j(j+1)\phi_{j,m}^{(\pm)} \ ,$$
$$\vec{L}\cdot\vec{\sigma}\phi_{j,m}^{(\pm)} \equiv (J^2 - L^2 - 3/4)\phi_{j,m}^{(\pm)} = -(k+1)\phi_{j,m}^{(\pm)} \ . \tag{5.84}$$

where k is an eigenvalue of the operator $-(\vec{\sigma}\cdot\vec{L}+1)$ with the eigenvalues $k = \pm1, \pm2, \pm3, ...$ and satisfies $\mid k \mid = (j+\frac{1}{2})$. In other words, $k = -(j+1/2)$

for the case $j = l + 1/2$ corresponding to the angular solution ϕ^+ and $k = j + 1/2$ when $j = l - 1/2$ corresponding to the angular solution ϕ^-. In terms of these the general solution to the central field problem can be written for a given j, m as

$$\psi_{jm} = \frac{1}{r} \begin{pmatrix} iG_k^+(r)\phi_{jm}^+ + iG_k^-(r)\phi_{jm}^- \\ F_k^+(r)\phi_{jm}^- + F_k^-(r)\phi_{jm}^+ \end{pmatrix} . \tag{5.85}$$

To compute the energy levels one only needs to concentrate on the radial equations which are given for example in the text of Bjorken and Drell:

$$G'(r) + \frac{kG}{r} - (\alpha_1 - V)F = 0 ,$$

$$F'(r) - \frac{kF}{r} - (\alpha_2 + V)G = 0 , \tag{5.86}$$

where

$$\alpha_1 = m + E , \quad \alpha_2 = m - E , \tag{5.87}$$

and $G_k(F_k)$ is the "large" ("small") component in the non-relativistic limit. The radial functions G_k and F_k must be multiplied by the appropriate two component angular eigenfunctions to make up the full four-component solutions of the Dirac equation as given in eq. (5.85). These coupled equations are in general not analytically solvable; one of the few exceptions being the case of the Dirac particle in a Coulomb field for which

$$V(r) = -\frac{\gamma}{r} , \quad \gamma = Ze^2 . \tag{5.88}$$

We now show that the Coulomb problem can also be solved algebraically by using the ideas of SUSY and shape invariance. To that end, we first note that in the case of the Coulomb potential, the coupled equations (5.86) can be written in a matrix form as

$$\begin{pmatrix} G'(r) \\ F'(r) \end{pmatrix} + \frac{1}{r} \begin{pmatrix} k & -\gamma \\ \gamma & -k \end{pmatrix} \begin{pmatrix} G \\ F \end{pmatrix} = \begin{pmatrix} 0 & \alpha_1 \\ \alpha_2 & 0 \end{pmatrix} \begin{pmatrix} G \\ F \end{pmatrix} . \tag{5.89}$$

Following Sukumar, we now notice that the matrix multiplying $1/r$ can be diagonalized by multiplying it by a matrix D from the left and D^{-1} from the right where

$$D = \begin{pmatrix} k + s & -\gamma \\ -\gamma & k + s \end{pmatrix} , \quad s = \sqrt{k^2 - \gamma^2} . \tag{5.90}$$

On multiplying eq. (5.89) from the left by the matrix D and introducing the new variable $\rho = Er$ leads to the pair of equations

$$\begin{aligned} A\tilde{F} &= (\frac{m}{E} - \frac{k}{s})\tilde{G} , \\ A^{\dagger}\tilde{G} &= -(\frac{m}{E} + \frac{k}{S})\tilde{F} , \end{aligned} \tag{5.91}$$

where

$$\begin{pmatrix} \tilde{G} \\ \tilde{F} \end{pmatrix} = D \begin{pmatrix} G \\ F \end{pmatrix} , \tag{5.92}$$

and

$$A = \frac{d}{d\rho} - \frac{s}{\rho} + \frac{\gamma}{s} , \quad A^{\dagger} = -\frac{d}{d\rho} - \frac{s}{\rho} + \frac{\gamma}{s} . \tag{5.93}$$

Thus we can easily decouple the equations for $\tilde{F}$ and $\tilde{G}$ thereby obtaining

$$\begin{aligned} H_1\tilde{F} \equiv A^{+}A\tilde{F} &= (\frac{k^2}{s^2} - \frac{m^2}{E^2})\tilde{F} , \\ H_2\tilde{G} \equiv AA^{+}\tilde{G} &= (\frac{k^2}{s^2} - \frac{m^2}{E^2})\tilde{G} . \end{aligned} \tag{5.94}$$

We thus see that there is a supersymmetry in the problem and $H_{1,2}$ are shape invariant supersymmetric partner potentials since

$$H_2(\rho; s, \gamma) = H_1(\rho; s + 1, \gamma) + \frac{\gamma^2}{s^2} - \frac{\gamma^2}{(s + 1)^2} . \tag{5.95}$$

On comparing with the formalism of Chap. 4 it is then clear that in this case

$$a_2 = s + 1 , \quad a_1 = s , \quad R(a_2) = \frac{\gamma^2}{a_1^2} - \frac{\gamma^2}{a_2^2} , \tag{5.96}$$

so that the energy eigenvalues of H_1 are given by

$$(\frac{k}{s})^2 - (\frac{m^2}{E_n^2}) \equiv E_n^{(1)} = \sum_{k=2}^{n+1} R(a_k) = \gamma^2(\frac{1}{s^2} - \frac{1}{(s + n)^2}) . \tag{5.97}$$

Thus the bound state energy eigenvalues E_n for Dirac particle in a Coulomb field are given by

$$E_n = \frac{m}{[1 + \frac{\gamma^2}{(s+n)^2}]^{1/2}} , \quad n = 0, 1, 2, \dots . \tag{5.98}$$

It should be noted that every eigenvalue of H_1 is also an eigenvalue of H_2 except for the ground state of H_1 which satisfies

$$A\tilde{F} = 0 \implies \tilde{F}_0(\rho) = \rho^s \exp\left(-\gamma\rho/s\right) . \tag{5.99}$$

Using the formalism for the SIP as developed in Chap. 4, one can also algebraically obtain all the eigenfunctions of $\tilde{F}$ and $\tilde{G}$.

Notice that the spectrum as given by eq. (5.98) only depends on $\mid k \mid$ leading to a doublet of states corresponding to $k = \mid k \mid$ and $k = - \mid k \mid$ for all positive n. However, for $n = 0$, only the negative value of k is allowed and hence this is a singlet state.

5.7   SUSY and the Dirac Particle in a Magnetic Field

Let us again consider the Dirac equation in an electromagnetic field as given by eq. (5.80) but now consider the other case when the vector potential is nonzero but the scalar potential is zero i.e. $A_0 = 0, \vec{A} \neq 0$. As shown earlier, the energy eigenvalues for this problem can be directly related to those of the Pauli equation and thus knowing those solutions we can also solve for the Dirac equation in a given magnetic field. To obtain the wave functions, however, it is useful to use a slightly different approach which comes to the same conclusions about the energy eigenvalues.

It was shown by Feynman and Gell-Mann and Laurie Brown that the solution of the four component Dirac equation in the presence of an external electromagnetic field can be generated from the solution of a two component relativistically invariant equation. In particular, if ψ obeys the two component equation

$$[(\vec{p} + \vec{A})^2 + m^2 + \vec{\sigma}\cdot(\vec{B} + i\vec{E})]\psi = (\overline{E} + A_0)^2\psi , \tag{5.100}$$

then the four component spinors that are solutions of the massive Dirac equation are generated from the two component ψ via

$$\psi_D = \begin{pmatrix} (\vec{\sigma}\cdot(\vec{p} + \vec{A}) + \overline{E} - A_0 + m)\psi \\ (\vec{\sigma}\cdot(\vec{p} + \vec{A}) + \overline{E} - A_0 - m)\psi \end{pmatrix} . \tag{5.101}$$

Thus, in order to solve the Dirac equation, it is sufficient to solve the much simpler two-component eq. (5.100) and then generate the corresponding Dirac solutions by the use of eq. (5.101). In the special case when the scalar potential A_0 (and hence $\vec{E}$) vanishes, the two-component equation then has the canonical form of the Pauli equation describing the motion of a charged particle in an external magnetic field. If further, $m = 0$ and the motion is confined to two dimensions, then the Pauli eq. (5.100) exactly reduces to eq. (5.25). Further, since

$$(H_D)^2 = [\vec{\alpha} \cdot (\vec{p} + \vec{A})]^2 = H_{Pauli} , \qquad (5.102)$$

hence, there is a supersymmetry in the massless Dirac problem in external magnetic fields in two dimensions as discussed earlier. We can now immediately borrow all the results of the section on the Pauli equation. In particular, it follows that if the total flux $\Phi(= \int B_z dx dy) = n + \epsilon (0 \le \epsilon < 1)$ then there are precisely $n - 1$ zero modes of the massless Dirac equation in two dimensions in the background of the external magnetic field B ($B \equiv B_z$). Further, in view of eqs. (5.100) and (5.101) we can immediately write down the exact solution of the massless Dirac equation in an external magnetic field in two dimensions in all the four situations discussed in Sec. 5.2 when the gauge potential depended on only one coordinate (say y). Further using the results of that section, one can also algebraically obtain the exact solution of the Dirac equation in a uniform magnetic field in the symmetric gauge when the gauge potential depends on both x and y.

Even though there is no SUSY, exact solutions of the Pauli and hence the Dirac equation are also possible in the massive case. On comparing the equations as given by (5.100) (with $A_0 = 0, \vec{E} = 0$) and (5.25) it is clear that the exact solutions in the massive case are simply obtained from the massless case by replacing $\overline{E}^2$ by $\overline{E}^2 - m^2$. Summarizing, we conclude that the exact solutions of the massive (as well as the massless) Dirac equation in an external magnetic field in two dimensions can be obtained algebraically if the magnetic field $B(\equiv B_Z)$ has any one of the following four forms

(1) $B = $ constant ,
(2) $B = -a \operatorname{sech}^2 y$,
(3) $B = -a \sec^2 y \ (-\pi/2 \le y \le \pi/2)$,
(4) $B = -c_2 \exp(-y)$.

Further, in the uniform magnetic field case, the solution can be obtained either in the asymmetric or in the symmetric gauge.

References

(1) H. Goldstein, *Classical Mechanics*, Addison-Wesley (1950).

(2) M. de Crombrugghe and V. Rittenberg, *Supersymmetric Quantum Mechanics*, Ann. Phys. (NY) **151** (1983) 99-126.

(3) G. Junker, *Supersymmetric Methods in Quantum and Statistical Physics*, Springer (1996).

(4) B. Thaller, *The Dirac Equation*, Springer Verlag (1992).

(5) F. Cooper, A. Khare, R. Musto, and A. Wipf, *Supersymmetry and the Dirac Equation*, Ann. Phys. **187** (1988) 1-28.

(6) J. Bjorken and S. Drell, *Relativistic Quantum Mechanics*, McGraw Hill (1964), Chapter 4.

(7) C. V. Sukumar, *Supersymmetry and the Dirac Equation for a Central Coulomb Field*, Jour. Phys. **A18** (1985) L697-L701.

Problems

1. Obtain the spectrum of the Pauli equation in case the gauge potential is

$$A_y(x,y) = 0 \ , \ A_x(x,y) = W(y) = A \tan \alpha y + c \ .$$

2. Obtain the energy eigenvalues of the Dirac equation in $1+1$ dimensions in case the Lorentz scalar potential $\phi(x)$ is

$$\phi(x) = a\epsilon(x) = a[\theta(x) - \theta(-x)] \ .$$

3. Obtain the band edge energies of the Dirac equation in $1+1$ dimensions in case the Lorentz scalar potential $\phi(x)$ is the periodic potential

$$\phi(x) = m\frac{\text{sn}\,x\ \text{cn}\,x}{\text{dn}\,x} \ .$$

where $\text{sn}\,x$, $\text{cn}\,x$, and $\text{dn}\,x$ are Jacobi elliptic functions.

4. Obtain the energy levels for the massless Dirac equation in a harmonic oscillator potential in $3+1$ dimensions when there is an equal admixture of both the scalar and the vector potentials.

Chapter 6

Isospectral Hamiltonians

In this chapter, we will describe how one can start from any given one-dimensional potential $V_1(x)$ with n bound states, and use supersymmetric quantum mechanics to construct an n-parameter family of strictly isospectral potentials $V_1(\lambda_1, \lambda_2, \ldots, \lambda_n; x)$ i.e., potentials with eigenvalues, reflection and transmission coefficients identical to those for $V_1(x)$. The fact that such families exist has been known for a long time from the inverse scattering approach, but the Gelfand-Levitan approach to finding them is technically much more complicated than the supersymmetry approach described here. Indeed, with the advent of SUSY QM, there is a revival of interest in the determination of isospectral potentials. In Sec. 6.1 we describe how a one parameter isospectral family is obtained by first deleting and then re-inserting the ground state of $V_1(x)$ using the Darboux procedure. The generalization to obtain an n-parameter family is described in Sec. 6.2. These isospectral families are closely connected to multi-soliton solutions of nonlinear integrable systems. In Sec. 6.3 we review the connection between inverse scattering theory and finding multisoliton solutions to nonlinear evolution equations. We then show that the n-parameter families of reflectionless isospectral potentials provide surprisingly simple expressions for the pure multi-soliton solutions of the Korteweg-de Vries (KdV) and other nonlinear evolution equations and thus provide a complementary approach to inverse scattering methods.

6.1   One Parameter Family of Isospectral Potentials

In this section, we describe two approaches for obtaining the one-parameter family $V_1(\lambda_1; x)$ of potentials isospectral to a given potential $V_1(x)$. One way of determining isospectral potentials is to consider the question of the uniqueness of the superpotential $W(x)$ in the definition of the partner potential to $V_1(x)$, namely $V_2(x)$. In other words, what are the various possible superpotentials $\hat{W}(x)$ other than $W(x)$ satisfying

$$V_2(x) = \hat{W}^2(x) + \hat{W}'(x) . \tag{6.1}$$

If there are new solutions, then one would obtain new potentials $\hat{V}_1(x) = \hat{W}^2 - \hat{W}'$ which would be isospectral to $V_1(x)$. To find the most general solution, let

$$\hat{W}(x) = W(x) + \phi(x) , \tag{6.2}$$

in eq. (6.1). We then find that $y(x) = \phi^{-1}(x)$ satisfies the Bernoulli equation

$$y'(x) = 1 + 2Wy , \tag{6.3}$$

whose solution is

$$\frac{1}{y(x)} = \phi(x) = \frac{d}{dx} \ln[\mathcal{I}_1(x) + \lambda_1] . \tag{6.4}$$

Here

$$\mathcal{I}_1(x) \equiv \int_{-\infty}^{x} \psi_1^2(x')dx' , \tag{6.5}$$

λ_1 is a constant of integration and $\psi_1(x)$ is the normalized ground state wave function of $V_1(x) = W^2(x) - W'(x)$. It may be noted here that unlike the rest of the book, in this chapter $\psi_1, \psi_2, \psi_3, \ldots$ denote the normalized ground state eigenfunctions of the isospectral family of potentials $V_1(x), V_2(x), V_3(x), \ldots$ respectively. Thus the most general $\hat{W}(x)$ satisfying eq. (6.1) is given by

$$\hat{W}(x) = W(x) + \frac{d}{dx} \ln[\mathcal{I}_1(x) + \lambda_1] , \tag{6.6}$$

so that all members of the one parameter family of potentials

$$\hat{V}_1(x) = \hat{W}^2(x) - \hat{W}'(x) = V_1(x) - 2\frac{d^2}{dx^2}\ln[\mathcal{I}_1(x) + \lambda_1] \,, \qquad (6.7)$$

have the same SUSY partner $V_2(x)$.

In the second approach, we delete the ground state ψ_1 at energy E_1 for the potential $V_1(x)$. This generates the SUSY partner potential $V_2(x) = V_1 - 2\frac{d^2}{dx^2}\ln\psi_1$, which has the same eigenvalues as $V_1(x)$ except for the bound state at energy E_1. The next step is to reinstate a bound state at energy E_1.

Although the potential V_2 does not have an eigenenergy E_1, the function $1/\psi_1$ satisfies the Schrödinger equation with potential V_2 and energy E_1. The other linearly independent solution is $\int_{-\infty}^{x}\psi_1^2(x')dx'/\psi_1$. Therefore, the most general solution of the Schrödinger equation for the potential V_2 at energy E_1 is

$$\Phi_1(\lambda_1) = (\mathcal{I}_1 + \lambda_1)/\psi_1 \,. \qquad (6.8)$$

Now, starting with a potential V_2, we can again use the standard SUSY (Darboux) procedure to add a state at E_1 by using the general solution $\Phi_1(\lambda_1)$,

$$\hat{V}_1(\lambda_1) = V_2 - 2\frac{d^2}{dx^2}\ln\Phi_1(\lambda_1) \,. \qquad (6.9)$$

The function $1/\Phi_1(\lambda_1)$ is the normalizable ground state wave function of $\hat{V}_1(\lambda_1)$, provided that λ_1 does not lie in the interval $-1 \leq \lambda_1 \leq 0$. Therefore, we find a one-parameter family of potentials $\hat{V}_1(\lambda_1)$ isospectral to V_1 for $\lambda_1 > 0$ or $\lambda_1 < -1$

$$\begin{aligned}
\hat{V}_1(\lambda_1) &= V_1 - 2\frac{d^2}{dx^2}\ln(\psi_1\Phi_1(\lambda_1)) \\
&= V_1 - 2\frac{d^2}{dx^2}\ln(\mathcal{I}_1 + \lambda_1) \,. \qquad (6.10)
\end{aligned}$$

The corresponding ground state wave functions are

$$\hat{\psi}_1(\lambda_1; x) = 1/\Phi_1(\lambda_1) \,. \qquad (6.11)$$

Note that this family contains the original potential V_1. This corresponds to the choices $\lambda_1 \to \pm\infty$.

To elucidate this discussion, it may be worthwhile to explicitly construct the one-parameter family of strictly isospectral potentials corresponding to the one dimensional harmonic oscillator. In this case

$$W(x) = \frac{\omega}{2}x , \tag{6.12}$$

so that

$$V_1(x) = \frac{\omega^2}{4}x^2 - \frac{\omega}{2} . \tag{6.13}$$

The normalized ground state eigenfunction of $V_1(x)$ is

$$\psi_1(x) = \left(\frac{\omega}{2\pi}\right)^{1/4} \exp(-\omega x^2/4) . \tag{6.14}$$

Using eq. (6.5) it is now straightforward to compute the corresponding $\mathcal{I}_1(x)$. We get

$$\mathcal{I}_1(x) = 1 - \frac{1}{2} \operatorname{erfc}\left(\frac{\sqrt{\omega}}{2}x\right) ; \quad \operatorname{erfc}(x) = \frac{2}{\sqrt{\pi}} \int_x^\infty e^{-t^2} dt . \tag{6.15}$$

Using eqs. (6.10) and (6.11), one obtains the one parameter family of isospectral potentials and the corresponding ground state wave functions. In Figs. 6.1 and 6.2 , we have plotted some of the potentials and the ground state wave functions for the case $\omega = 2$.

We see that as λ_1 decreases from ∞ to 0, $\hat{V}_1$ starts developing a minimum which shifts towards $x = -\infty$. Note that as λ_1 finally becomes zero this attractive potential well is lost and we lose a bound state. The remaining potential is called the Pursey potential $V_P(x)$. The general formula for $V_P(x)$ is obtained by putting $\lambda_1 = 0$ in eq. (6.10). An analogous situation occurs in the limit $\lambda_1 = -1$, the remaining potential being the Abraham-Moses potential.

6.2   Generalization to n-Parameter Isospectral Family

The second approach discussed in the previous section can be generalized by first deleting all n bound states of the original potential $V_1(x)$ and then reinstating them one at a time. Since one parameter is generated every time an eigenstate is reinstated, the final result is a n-parameter isospectral family. Recall that deleting the eigenenergy E_1 gave the potential $V_2(x)$. The ground state ψ_2 for the potential V_2 is located at energy E_2

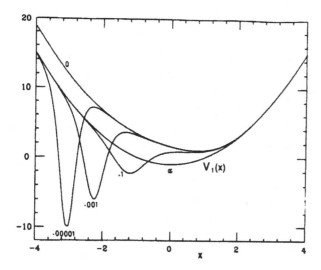

Fig. 6.1   Selected members of the family of potentials with energy spectra identical to the one dimensional harmonic oscillator with $\omega = 2$. The choice of units is $\hbar = 2m = 1$. The curves are labeled by the value of λ_1, and cover the range $0 < \lambda_1 \leq \infty$. The curve $\lambda_1 = \infty$ is the one dimensional harmonic oscillator. The curve marked $\lambda_1 = 0$ is known as the Pursey potential and has one bound state less than the oscillator.

The procedure can be repeated "upward", producing potentials $V_3, V_4, \ldots$ with ground states $\psi_3, \psi_4, \ldots$ at energies $E_3, E_4, \ldots$, until the top potential $V_{n+1}(x)$ holds no bound state (see Fig. 6.3 , which corresponds to $n = 2$).

In order to produce a two-parameter family of isospectral potentials, we go from V_1 to V_2 to V_3 by successively deleting the two lowest states of V_1 and then we re-add the two states at E_2 and E_1 by SUSY transformations. The most general solutions of the Schrödinger equation for the potential V_3 are given by $\Phi_2(\lambda_2) = (\mathcal{I}_2 + \lambda_2)/\psi_2$ at energy E_2, and $A_2\Phi_1(\lambda_1)$ at energy E_1 (see Fig. 6.3). The quantities $\mathcal{I}_i$ are defined by

$$\mathcal{I}_i(x) \equiv \int_{-\infty}^{x} \psi_i^2(x')dx' . \tag{6.16}$$

Here the SUSY operator A_i relates solutions for the potentials V_i and V_{i+1},

$$A_i = \frac{d}{dx} - (\ln \psi_i)' . \tag{6.17}$$

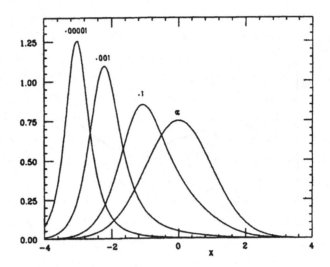

Fig. 6.2   Ground state wave functions for all the potentials shown in Fig. 6.1, except the Pursey potential.

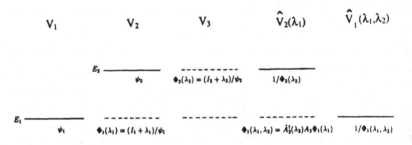

Fig. 6.3   A schematic diagram showing how SUSY transformations are used for deleting the two lowest states of a potential $V_1(x)$ and then re-inserting them, thus producing a two-parameter (λ_1, λ_2) family of potentials isospectral to $V_1(x)$.

Then, as before, we find an isospectral one-parameter family $\hat{V}_2(\lambda_2)$,

$$\hat{V}_2(\lambda_2) = V_2 - 2\frac{d^2}{dx^2}\ln(\mathcal{I}_2 + \lambda_2) \ . \tag{6.18}$$

The solutions of the Schrödinger equation for potentials V_3 and $\hat{V}_2(\lambda_2)$ are related by a new SUSY operator

$$\hat{A}_2^\dagger(\lambda_2) = -\frac{d}{dx} + (\ln \Phi_2(\lambda_2))' \ . \tag{6.19}$$

Therefore, the solution $\Phi_1(\lambda_1, \lambda_2)$ at E_1 for $\hat{V}_2(\lambda_2)$ is

$$\Phi_1(\lambda_1, \lambda_2) = \hat{A}_2^\dagger(\lambda_2) A_2 \Phi_1(\lambda_1) \ . \tag{6.20}$$

The normalizable function $1/\Phi_1(\lambda_1, \lambda_2)$ is the ground state at E_1 of a new potential, which results in a two-parameter family of isospectral systems $\hat{V}_1(\lambda_1, \lambda_2)$,

$$
\begin{aligned}
\hat{V}_1(\lambda_1, \lambda_2) &= V_1 - 2\frac{d^2}{dx^2}\ln(\psi_1\psi_2\Phi_2(\lambda_2)\Phi_1(\lambda_1, \lambda_2)) \\
&= V_1 - 2\frac{d^2}{dx^2}\ln(\psi_1(\mathcal{I}_2 + \lambda_2)\Phi_1(\lambda_1, \lambda_2)) \ , \tag{6.21}
\end{aligned}
$$

for $\lambda_i > 0$ or $\lambda_i < -1$. A useful alternative expression is

$$\hat{V}_1(\lambda_1, \lambda_2) = -\hat{V}_2(\lambda_2) + 2(\Phi_1'(\lambda_1, \lambda_2)/\Phi_1(\lambda_1, \lambda_2))^2 + 2E_1 \ . \tag{6.22}$$

The above procedure is best illustrated by the pyramid structure in Fig. 6.3. It can be generalized to an n-parameter family of isospectral potentials for an initial system with n bound states. The formulas for an n-parameter family are

$$\Phi_i(\lambda_i) = (\mathcal{I}_i + \lambda_i)/\psi_i \ ; \quad i = 1, \cdots, n \ , \tag{6.23}$$

$$A_i = \frac{d}{dx} - (\ln\psi_i)' \ , \tag{6.24}$$

$$\hat{A}_i^\dagger(\lambda_i, \cdots, \lambda_n) = -\frac{d}{dx} + [\ln\Phi_i(\lambda_i, \cdots, \lambda_n)]' \ , \tag{6.25}$$

$$
\begin{aligned}
&\Phi_i(\lambda_i, \lambda_{i+1}, \cdots, \lambda_n) \\
&= \hat{A}_{i+1}^\dagger(\lambda_{i+1}, \lambda_{i+2}, \cdots, \lambda_n)\hat{A}_{i+2}^\dagger(\lambda_{i+2}, \lambda_{i+3}, \cdots, \lambda_n) \cdots \hat{A}_n^\dagger(\lambda_n) \\
&\times \ A_n A_{n-1} \cdots A_{i+1}\Phi_i(\lambda_i) \ , \tag{6.26}
\end{aligned}
$$

$$\hat{V}_1(\lambda_1, \cdots, \lambda_n) = V_1 - 2\frac{d^2}{dx^2}\ln(\psi_1\psi_2\cdots\psi_n\Phi_n(\lambda_n)\cdots\Phi_1(\lambda_1, \cdots, \lambda_n)) \ . \tag{6.27}$$

The above equations summarize the main results of this section.

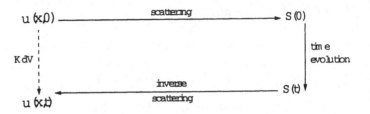

Fig. 6.4   Flow chart showing the connection between inverse scattering and solution of the KdV equation.

6.3   Inverse Scattering and Solitons

We would like to apply the formalism for isospectral Hamiltonians just developed above to obtain multisoliton solutions of the KdV equation. Before embarking on this it is useful to review the main ideas relating inverse scattering methods and soliton solutions.

It is interesting that the flow equations related to completely integrable dynamical systems, such as the Korteweg-de Vries equation can be exactly solved by solving a related one dimensional quantum mechanical problem. To be specific, if we consider the KdV equation

$$u_\tau - 6uu_x + u_{xxx} = 0 , \quad \tau > 0 , \tag{6.28}$$

with $u(x,0) = f(x)$, this defines a particular evolution in the parameter τ. If we consider a time independent Schrödinger equation which also depends on the parameter τ (which is not to be confused with the time t in the time dependent Schrödinger equation)

$$\psi_{xx}(x,\tau) + (\lambda - u(x,\tau))\psi(x,\tau) = 0 , \tag{6.29}$$

it is possible to show, that the bound state energy eigenvalues $\lambda = -\kappa_n^2$ are independent of the parameter τ if $u(x,\tau)$ obeys the KdV equation. Thus to find these eigenvalues one only needs $f(x)$. If we now know how the wave function ψ flows in the parameter τ in the limits $x \to \pm\infty$, we can then reconstruct $u(x,\tau)$ from the inverse scattering problem in terms of the scattering data at arbitrary τ. This strategy is summarized in Fig.6.4 where $S(t)$ denotes the scattering data $R(k,t), \kappa(t)$ and $c_n(t)$ defined below.

First let us summarize the main results of inverse scattering theory. The derivations can, for example, be found in the book by Drazin and Johnson. Here we will assume unlike our earlier convention, that a particle is incident on a potential from the right to conform with Drazin and Johnson. Also in the soliton literature, the ground state wave function and energy is denoted by by ψ_1, E_1 instead of ψ_0, E_0 so to conform with that literature we will use this altered convention in what follows.

The Schrödinger equation we want to solve is written as:

$$-\psi_{xx} + u\psi = \lambda\psi .$$ (6.30)

The potentials we are interested in have the property $u(x) \to 0$ as $x \to \pm\infty$. Thus for the continuous spectrum we have in the asymptotic regime:

$$\psi(x,k) \sim \left\{ \begin{array}{ll} e^{-ikx} + R(k)e^{ikx} & \text{as } x \to +\infty \\ T(k)e^{-ikx} & \text{as } x \to -\infty \end{array} \right\}$$ (6.31)

for $\lambda = E = k^2 > 0$. For the bound state spectra we have instead:

$$\psi^{(n)}(x) \sim c_n e^{-\kappa_n x} \text{ as } x \to +\infty ,$$ (6.32)

where now $\lambda = E_n = -\kappa_n^2 < 0$, for each discrete eigenvalue ($n = 1, 2, \cdots, N$). Note the special notation for the n bound states here ordered by the asymptotic behavior. It can be shown that in terms of c_n, κ_n and the reflection coefficients $R(k)$ one can reconstruct the potential $u(x)$ in the following manner. Defining the function

$$F(X) = \sum_{n=1}^{N} c_n^2 e^{-\kappa_n X} + \frac{1}{2\pi} \int_{-\infty}^{\infty} R(k)e^{ikX} dk ,$$ (6.33)

one then constructs a new function $K(x,z)$ which is the solution of the Marchenko equation :

$$K(x,z) + F(x+z) + \int_{x}^{\infty} K(x,y)F(y+z)dy = 0 .$$ (6.34)

In terms of $K(x,z)$ one finds:

$$u(x) = -2\frac{dK(x,x)}{dx} .$$ (6.35)

So now thinking of the time independent Schrödinger equation as a flow equation for $\psi(x,\tau)$ and differentiating it, assuming that $u(x,\tau)$ obeys the flow equation for the KdV equation, one then finds that the discrete eigenvalues,

$$E_n = -\kappa_n^2$$

are independent of τ. Furthermore, one can show that the c_n and $R(k)$ and $T(k)$ obey the following flow equations:

$$\frac{dc_n}{d\tau} - 4\kappa_n^3 c_n = 0 ; \quad \rightarrow c_n(\tau) = c_n(0)e^{4\kappa_n^3 \tau} ,$$

$$\frac{dR(k,\tau)}{d\tau} - 8ik^3 R(k,\tau) = 0 ; \quad \rightarrow R(k,\tau) = R(k,0)e^{8ik^3 \tau} ,$$

$$\frac{dT(k,\tau)}{d\tau} = 0 , \quad \rightarrow T(k,\tau) = T(k,0) . \tag{6.36}$$

It is clear that the flow evolution will get more complicated as we choose $f(x) = u(x,0)$ to correspond to having more and more bound states because of the Marchenko equation. For example if we start with a solvable shape invariant potential with only one bound state and which corresponds to a reflectionless potential (see Chap. 4)

$$f(x) = u(x,0) = -2 \operatorname{sech}^2(x) , \tag{6.37}$$

then there is one normalized bound state with $\kappa = 1$:

$$\psi^{(1)}(x) = \frac{1}{\sqrt{2}} \operatorname{sech} x \sim \sqrt{2}e^{-x} \text{ as } x \to \infty , \tag{6.38}$$

so that $c_1(0) = \sqrt{2}$ and $c_1(\tau) = \sqrt{2}e^{4\tau}$. Because of the reflectionless nature of the potential, $R(k) = 0$ making it easy to solve the Marchenko equation and one obtains:

$$u(x,\tau) = -2 \operatorname{sech}^2(x - 4\tau) . \tag{6.39}$$

If we now take an initial condition where there are exactly two bound states and which again correspond to a reflectionless potential:

$$u(x,0) = -6 \operatorname{sech}^2 x ,$$

one finds for the bound state wave functions at $\tau = 0$, having $\kappa_1 = 2$, and $\kappa_2 = 1$

$$\psi^{(1)}(x) = \frac{\sqrt{3}}{2}\operatorname{sech}^2 x ; \quad \psi^{(2)}(x) = \frac{\sqrt{3}}{\sqrt{2}} \tanh x \operatorname{sech} x . \tag{6.40}$$

From the asymptotic behavior and the flow equation one then finds:

$$c_1(\tau) = 2\sqrt{3}e^{32\tau} ; \quad c_2(\tau) = \sqrt{6}e^{4\tau} . \tag{6.41}$$

Using the machinery of the inverse scattering formalism, one eventually finds

$$u(x,\tau) = -12\,\frac{3 + 4\cosh(2x - 8\tau) + \cosh(4x - 64\tau)}{\{3\cosh(x - 28\tau) + \cosh(3x - 36\tau)\}^2} \,. \qquad (6.42)$$

This is the form of the two soliton solution at arbitrary evolution time τ. We shall now rederive this solution using the method of isospectral Hamiltonians.

As an application of isospectral potential families, we consider reflectionless potentials of the form

$$V_1 = -n(n + 1)\mathrm{sech}^2 x \,, \qquad (6.43)$$

where n is an integer, since these potentials are of special physical interest. V_1 holds n bound states, and we may form a n-parameter family of isospectral potentials. We start with the simplest case $n = 1$. We have $V_1 = -2\mathrm{sech}^2 x, E_1 = -1$ and $\psi_1 = \frac{1}{\sqrt{2}}\,\mathrm{sech}\,x$. The corresponding 1-parameter family is

$$\hat{V}_1(\lambda_1) = -2\,\mathrm{sech}^2(x + \frac{1}{2}\ln[1 + \frac{1}{\lambda_1}]) \,. \qquad (6.44)$$

Clearly, varying the parameter λ_1 corresponds to translations of $V_1(x)$. As λ_1 approaches the limits 0^+ (Pursey limit) and -1^- (Abraham-Moses limit), the minimum of the potential moves to $-\infty$ and $+\infty$ respectively.

For the case $n = 2$, $V_1 = -6\mathrm{sech}^2 x$ and there are two bound states at $E_1 = -4$ and $E_2 = -1$. The SUSY partner potential is $V_2 = -2\,\mathrm{sech}^2 x$. The ground state wave functions of V_1 and V_2 are $\psi_1 = \frac{\sqrt{3}}{2}\mathrm{sech}^2 x$ and $\psi_2 = \frac{1}{\sqrt{2}}\mathrm{sech}\,x$. Also, $\mathcal{I}_1 = \frac{1}{4}(\tanh x + 1)^2(2 - \tanh x)$ and $\mathcal{I}_2 = \frac{1}{2}(\tanh x + 1)$. After some algebraic work, we obtain the 2-parameter family

$$\hat{V}_1(\lambda_1, \lambda_2) = -12\frac{[3 + 4\cosh(2x - 2\delta_2) + \cosh(4x - 2\delta_1)]}{[\cosh(3x - \delta_2 - \delta_1) + 3\cosh(x + \delta_2 - \delta_1)]^2} \,,$$

$$\delta_i \equiv -\frac{1}{2}\ln(1 + \frac{1}{\lambda_i}) \,, \quad i = 1, 2 \,.$$

As we let $\lambda_1 \to -1$, a well with one bound state at E_1 will move in the $+x$ direction leaving behind a shallow well with one bound state at E_2. The movement of the shallow well is essentially controlled by the parameter λ_2. Thus, we have the freedom to move either of the wells.

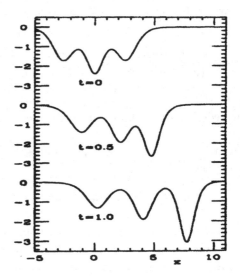

Fig. 6.5   The pure three-soliton solution of the KdV equation as a function of position (x) and time (t). This solution results from constructing the isospectral potential family starting from a reflectionless, symmetric potential with bound states at energies $E_1 = -25/16, E_2 = -1, E_3 = -16/25$.

In case we choose δ_1, δ_2 to be 32 and 4 respectively then we find that this solution is identical to the two soliton solution (6.42) as obtained from the inverse scattering formalism.

It is tedious but straightforward to obtain the result for arbitrary n and get $\hat{V}_1(\lambda_1, \lambda_2, \cdots, \lambda_n, x)$. It is well known that one-parameter (t) families of isospectral potentials can also be obtained as solutions of a certain class of nonlinear evolution equations. These equations have the form ($q = 0, 1, 2, \cdots$)

$$-u_t = (L_u)^q \, u_x \, , \qquad (6.45)$$

where the operator L_u is defined by

$$L_u f(x) = f_{xx} - 4uf + 2u_x \int_x^\infty dy f(y) \, , \qquad (6.46)$$

and u is chosen to vanish at infinity. (For $q = 0$ we simply get $-u_t = u_x$, while for $q = 1$ we obtain the well studied Korteweg-de Vries (KdV) equation). These equations are also known to possess pure (i.e., reflectionless) multisoliton solutions. It is possible to show that by suitably choosing the parameters λ_i as functions of t in the n-parameter SUSY isospectral family of a symmetric reflectionless potential holding n bound states, we can obtain an explicit analytic formula for the n-soliton solution of each of the above evolution equations. These expressions for the multisoliton solutions of eq. (6.45) are much simpler than any previously obtained using other procedures. Nevertheless, rather than displaying the explicit algebraic expressions here, we shall simply illustrate the three soliton solution of the KdV equation. The potentials shown in Fig. 6.5 are all isospectral and reflectionless holding bound states at $E_1 = -25/16$, $E_2 = -1, E_3 = -16/25$. As t increases, note the clear emergence of the three independent solitons.

In this section, we have found n-parameter isospectral families by repeatedly using the supersymmetric Darboux procedure for removing and inserting bound states. However, as briefly mentioned in Sec. 6.1, there are two other closely related, well established procedures for deleting and adding bound states. These are the Abraham-Moses procedure and the Pursey procedure. If these alternative procedures are used, one gets new potential families all having the same bound state energies but different reflection and transmission coefficients.

References

(1) G. Darboux, *Sur une Proposition Relative aux Équations Linéaires*, Comptes Rendu Acad. Sci. (Paris) **94** (1882) 1456-1459.

(2) C. Gardner, J. Greene, M. Kruskal and R. Miura, *Korteweg-de Vries Equation and Generalizations*, Comm. Pure App. Math. **27** (1974) 97-133.

(3) W. Eckhaus and A. Van Harten, *The Inverse Scattering Transformation and the Theory of Solitons*, North-Holland (1981).

(4) G.L. Lamb, *Elements of Soliton Theory*, John Wiley (1980).

(5) A. Das, *Integrable Models*, World Scientific (1989).

(6) P.G. Drazin and R.S. Johnson, *Solitons: an Introduction*, Cambridge University Press (1989).

(7) M.M. Nieto, *Relationship Between Supersymmetry and the Inverse Method in Quantum Mechanics*, Phys. Lett. **B145** (1984) 208-210.

(8) P.B. Abraham and H.E. Moses, *Changes in Potentials due to Changes in the Point Spectrum: Anharmonic Oscillators with Exact Solutions*, Phys. Rev. **A22** (1980) 1333-1340.

(9) D.L. Pursey, *New Families of Isospectral Hamiltonians*, Phys. Rev. **D33** (1986) 1048-1055.

(10) C.V. Sukumar, *Supersymmetric Quantum Mechanics of One Dimensional Systems*, J. Phys. **A18** (1985) 2917-2936; *Supersymmetric Quantum Mechanics and the Inverse Scattering Method*, ibid **A18** (1985) 2937-2955; *Supersymmetry, Potentials With Bound States at Arbitrary Energies and Multi-Soliton Configurations*, ibid **A19** (1986) 2297-2316.

(11) A. Khare and U.P. Sukhatme, *Phase Equivalent Potentials Obtained From Supersymmetry*, J. Phys. **A22** (1989) 2847-2860.

(12) W.-Y. Keung, U. Sukhatme, Q. Wang and T. Imbo, *Families of Strictly Isospectral Potentials*, J. Phys. **A22** (1989) L987-L992.

(13) Q. Wang, U. Sukhatme, W.-Y. Keung and T. Imbo, *Solitons From Supersymmetry*, Mod. Phys. Lett. **A5** (1990) 525-530.

Problems

1. Work out the one parameter family of potentials which are strictly isospectral to the infinite square well. Write down the ground state eigenfunction for these potentials, along with an explicit expression for computing all excited state eigenfunctions.

2. Show that the one parameter family of isospectral potentials coming from the potential $V_1(x) = 1 - 2 \operatorname{sech}^2 x$ is given by $V_1(x, \lambda) = 1 - 2 \operatorname{sech}^2(x + a)$ and prove that the constants a and λ are related by $a = \frac{1}{2} \ln(1 + \lambda^{-1})$.

3. Let $V_1(x)$ be a symmetric potential with normalized ground state wave function $\psi_1(x)$. Prove that if the potential $V_1(x, \lambda)$ belongs to the isospectral family of $V_1(x)$, then so does the parity reflected potential $V_1(-x, \lambda)$.

4. Work out the one parameter family of potentials which are strictly isospectral to the potential $V_1(x) = -6 \operatorname{sech}^2 x$. Write down the ground state wave function for any member of this family of potentials.

5. Compute the traveling-wave solutions [in the form $u(x,t) = f(x - ct)$] of the following three nonlinear evolution equations:

(i) Burgers equation $u_t = u_{xx} - uu_x$ with $u \to 0, x \to +\infty$, $u \to u_0, x \to -\infty$;

(ii) KdV equation $u_t = 6uu_x - u_{xxx}$ with $u, u_x, u_{xx} \to 0, x \to \pm\infty$;

(iii) Modified KdV equation $u_t = -6u^2 u_x - u_{xxx}$ with $u, u_x, u_{xx} \to 0, x \to \pm\infty$.

Chapter 7

New Periodic Potentials from Supersymmetry

So far we have considered potentials which have discrete and/or continuum spectra and by using SUSY QM methods we have generated new solvable potentials. In this section we extend this discussion to periodic potentials and their band spectra. The importance of this problem can hardly be overemphasized. For example, the energy spectrum of electrons on a lattice is of central importance in condensed matter physics. In particular, knowledge of the existence and locations of band edges and band gaps determines many physical properties of these systems. Unfortunately, even in one dimension, there are very few analytically solvable periodic potential problems. We show in this chapter that SUSY QM allows us to enlarge this class of solvable periodic potential problems. We will also discuss here some quasi-exactly solvable periodic potentials. Further, we will show that for periodic potentials, even though SUSY is unbroken, the Witten index can be zero.

7.1 Unbroken SUSY and the Value of the Witten Index

We start from the Hamiltonians $H_{1,2}$ in which the SUSY partner potentials $V_{1,2}$ are periodic nonsingular potentials with a period L. In view of the periodicity, one seeks solutions of the Schrödinger equation subject to the Bloch condition

$$\psi(x + L) = e^{ikL}\psi(x),\qquad(7.1)$$

where k is real and denotes the crystal momentum. As a result, the spectrum shows energy bands whose edges correspond to $kL = 0, \pi$, that is, the wave function at the band edges satisfy $\psi(x + L) = \pm\psi(x)$. For periodic potentials, the band edge energies and wave functions are often called eigenvalues and eigenfunctions, and we will use this terminology in this book. In particular the ground state eigenvalue and eigenfunction refers to the bottom edge of the lowest energy band.

Let us first discuss the question of SUSY breaking for periodic potentials. Since $H_1 = A^\dagger A$ and $H_2 = A A^\dagger$ are formally positive operators their spectrum is nonnegative and almost the same. The caveat "almost" is needed because the mapping between the positive energy states of the two does not apply to zero energy states.

The Schrödinger equation for $H_{1,2}$ has zero energy modes given by

$$\psi_0^{(1,2)}(x) = \exp\left(\mp \int^x dy W(y) \right), \qquad (7.2)$$

provided $\psi_0^{(1,2)}$ belong to the Hilbert space. Supersymmetry is unbroken if at least one of the $\psi_0^{(1,2)}$ is a true zero mode while otherwise it is dynamically broken. Thus in the broken case, the spectra of $H_{1,2}$ are identical and there are no zero modes. For a non periodic potential we have seen that at most one of the functions $\psi_0^{(1,2)}$ can be normalizable and hence an acceptable eigenfunction. By convention we are choosing W such that only H_1 (if at all) has a zero mode.

Let us now consider the case when W (and hence $V_{1,2}$) are periodic with period L. Now the eigenfunctions including the ground state wave function must satisfy the Bloch condition (7.1). But, in view of eq. (7.2) we have

$$\psi_0^{(1,2)}(x + L) = e^{\pm\phi_L}\psi_0^{(1,2)}(x), \qquad (7.3)$$

where

$$\phi_L = \int_x^{x+L} W(y)dy . \qquad (7.4)$$

On comparing eqs. (7.1) and (7.3) it is clear that for either of the wave functions $\psi_0^{(1,2)}$ to belong to the Hilbert space, we must identify $\pm\phi_L = ikL$. But ϕ_L is real (since W and hence $V_{1,2}$ are assumed to be real), which means that $\phi_L = 0$. Thus, the two functions $\psi_0^{(1,2)}$ either *both* belong to the Hilbert space, in which case they are strictly periodic with period L:

$\psi_0^{(1,2)}(x+L) = \psi_0^{(1,2)}(x)$, or (when $\phi_L \neq 0$) *neither of them* belongs to the Hilbert space. Thus in the periodic case, irrespective of whether SUSY is broken or unbroken, the spectra of $V_{1,2}$ is always strictly isospectral.

To summarize, we see that

$$\phi_L = \int_0^L W(y)dy = 0 \tag{7.5}$$

is a necessary condition for unbroken SUSY, and when this condition is satisfied then $H_{1,2}$ have identical spectra, including zero modes. In this case, using the known eigenfunctions $\psi_n^{(1)}(x)$ of $V_1(x)$ one can immediately write down the corresponding (un-normalized) eigenfunctions $\psi_n^{(2)}(x)$ of $V_2(x)$. In particular, from eq. (7.3) the ground state of $V_2(x)$ is given by

$$\psi_0^{(2)}(x) = \frac{1}{\psi_0^{(1)}(x)} = e^{\int^x W(y)\,dy} \,, \tag{7.6}$$

while the excited states $\psi_n^{(2)}(x)$ are obtained from $\psi_n^{(1)}(x)$ by using the relation

$$\psi_n^{(2)}(x) = [\frac{d}{dx} + W(x)]\psi_n^{(1)}(x) \,, \ (n \geq 1) \,. \tag{7.7}$$

Thus by starting from an exactly solvable periodic potential $V_1(x)$, one gets another strictly isospectral periodic potential $V_2(x)$.

We recall from Chap. 3 that the Witten index, $\Delta = \mathrm{Tr}(-1)^F = n_1 - n_2$ counts the difference between the number of zero modes $\psi_0^{(1)}$ and $\psi_0^{(2)}$ and is an indicator of SUSY breaking. In particular, if $\Delta \neq 0$ then there must be at least one zero mode and so SUSY is unbroken. On the other hand, if $\Delta = 0$ then more information is needed about whether both or neither partner potential has a zero energy state. Much of the power of the index method comes from the fact that the Witten index can be calculated quite easily and reliably both in SUSY QM and in SUSY field theory. This is because, to a large extent Δ is independent of the parameters (like masses, couplings, volume etc.) of the theory. The remarkable thing for periodic potentials is that when condition (7.5) is satisfied then SUSY is unbroken and yet the Witten index is always zero since both $H_{1,2}$ have equal number of zero modes.

As an illustration, consider $W(x) = A\cos x + B\sin 2x$. In this case $L = 2\pi$ and ϕ_L as given by eq. (7.5) is indeed zero so that the two partner potentials $V_{1,2}(x)$ have identical spectra including zero modes. The Witten index is therefore zero even though SUSY is still unbroken.

At this stage, it is worth pointing out that there are some special classes of periodic superpotentials which trivially satisfy the condition (7.5) and hence for them SUSY is unbroken. For example, suppose the superpotential is antisymmetric on a half-period:

$$W(x + \frac{L}{2}) = -W(x). \tag{7.8}$$

Then,

$$V_{1,2}(x + \frac{L}{2}) \equiv W^2(x + \frac{L}{2}) \mp W(x + \frac{L}{2}) = V_{2,1}(x). \tag{7.9}$$

Thus in this case $V_{1,2}$ are simply translations of one another by half a period, and hence are essentially identical in shape. Therefore, they must support exactly the same spectrum, as SUSY indeed tells us they do. Such a pair of isospectral $V_{1,2}$ that are identical in shape are termed as "self-isospectral". A simple example of a superpotential of this type is $W(x) = \cos x$, so that $V_2(x) = \cos^2 x - \sin x = V_1(x + \pi)$. In a way, self-isospectral potentials are uninteresting since in this case, SUSY will give us nothing new.

More generally, if a pair of periodic partner potentials $V_{1,2}$ are such that V_2 is just the partner potential V_1 up to a discrete transformation- a translation by any constant amount, a reflection, or both, then such a pair of partner potentials are termed as "self-isospectral ". For example, consider periodic superpotentials that are even functions of x:

$$W(-x) = W(x), \tag{7.10}$$

but which also satisfy the condition (7.5). Since the function $dW(x)/dx$ is now odd hence it follows that

$$V_{1,2}(-x) = V_{2,1}(x). \tag{7.11}$$

The partner potentials are then simply reflections of one another. They therefore have the same shape and hence give rise to exactly the same spectrum. A simple example of a superpotential of this type is again $W(x) = \cos x$, so that $V_2(x) = \cos^2 x - \sin x = V_1(-x)$.

It must be made clear here that not all periodic partner potentials are self-isospectral even though they are strictly isospectral. Consider for example, periodic superpotentials that are odd functions of x:

$$W(-x) = -W(x). \tag{7.12}$$

Then the condition (7.5) is satisfied trivially and hence SUSY is unbroken even though Witten index is zero. The function $dW(x)/dx$ is even and thus $V_{1,2}(x)$ are also even. In this case, $V_{\pm}(x)$ are not necessarily related by simple translations or reflections. For example, the superpotential $W(x) = A \sin x + B \sin 2x$ gives rise to an isospectral pair which is not self-isospectral. On the other hand, $W(x) = A \sin x + B \sin 3x$ gives rise to a self-isospectral pair since this W satisfies the condition (7.8).

7.2 Lamé Potentials and Their Supersymmetric Partners

The classic text book example of a periodic potential which is often used to demonstrate band structure is the Kronig-Penney model,

$$V(x) = \sum_{-\infty}^{\infty} V_0 \delta(x - nL). \tag{7.13}$$

It should be noted that the band edges of this model can only be computed by solving a transcendental equation.

Another well studied class of periodic problems consists of the Lamé potentials

$$V(x,m) = p\, m\, \mathrm{sn}^2(x,m), \quad p \equiv a(a+1). \tag{7.14}$$

Here $\mathrm{sn}(x,m)$ is a Jacobi elliptic function of real elliptic modulus parameter $m(0 \le m \le 1)$ with period $4K(m)$, where $K(m)$ is the " real elliptic quarter period " given by

$$K(m) = \int_0^{\frac{\pi}{2}} \frac{d\theta}{\sqrt{1 - m \sin^2 \theta}}. \tag{7.15}$$

For simplicity, from now on, we will not explicitly display the modulus parameter m as an argument of Jacobi elliptic functions unless necessary. Note that the elliptic function potentials (7.14) have period $2K(m)$. They will be referred to as Lamé potentials, since the corresponding Schrödinger equation is called the Lamé equation in the mathematics literature. It is known that for any integer value $a = 1, 2, 3, ...$, the corresponding Lamé potential has a bound bands followed by a continuum band. All the band edge energies and the corresponding wave functions are analytically known. We shall now apply the formalism of SUSY QM and calculate the SUSY partner potentials corresponding to the Lamé potentials as given by eq. (7.14) and show that even though $a = 1$ Lamé partners are self-isospectral, for $a \geq 2$ they are not self-isospectral. Consequently, SUSY QM generates new exactly solvable periodic problems!

Before we start our discussion, it is worth mentioning a few basic properties of the Jacobi elliptic functions $\mathrm{sn}\, x, \mathrm{cn}\, x$ and $\mathrm{dn}\, x$ which we shall be using in this discussion. First of all, whereas $\mathrm{sn}\, x$ and $\mathrm{cn}\, x$ have period $4K(m)$, $\mathrm{dn}\, x$ has period $2K(m)$ (i.e. $\mathrm{dn}(x + 2K(m)) = \mathrm{dn}\, x$). They are related to each other by

$$m\, \mathrm{sn}^2 x = m - m\, \mathrm{cn}^2 x = 1 - \mathrm{dn}^2 x \,. \qquad (7.16)$$

Further,

$$\frac{d}{dx}\mathrm{sn}\, x = \mathrm{cn}\, x\, \mathrm{dn}\, x\,; \frac{d}{dx}\mathrm{cn}\, x = -\mathrm{sn}\, x\, \mathrm{dn}\, x\,, \frac{d}{dx}\mathrm{dn}\, x = -m\, \mathrm{sn}\, x\, \mathrm{cn}\, x\,. \quad (7.17)$$

Besides

$$\mathrm{sn}(x + K) = \frac{\mathrm{cn}\, x}{\mathrm{dn}\, x}\,; \ \mathrm{cn}(x + K) = -\sqrt{1 - m}\frac{\mathrm{sn}\, x}{\mathrm{dn}\, x}\,; \ \mathrm{dn}(x + K) = \frac{\sqrt{1 - m}}{\mathrm{dn}\, x}\,.$$
$$(7.18)$$

Finally, for $m = 1(0)$, these functions reduce to the familiar hyperbolic (trigonometric) functions, i.e.

$$\mathrm{sn}(x, m = 1) = \tanh x\,; \mathrm{cn}(x, m = 1) = \mathrm{sech}\, x\,; \mathrm{dn}(x, m = 1) = \mathrm{sech}\, x\,,$$
$$\mathrm{sn}(x, m = 0) = \sin x\,; \mathrm{cn}(x, m = 0) = \cos x\,; \mathrm{dn}(x, m = 0) = 1\,. \quad (7.19)$$

Let us notice that when $m = 1$, the Lamé potentials (7.14) reduce to the well known Pöschl-Teller potentials

$$V(x, m = 1) = a(a + 1) - a(a + 1)\mathrm{sech}^2 x\,, \qquad (7.20)$$

which for integer a are known to be reflectionless and to have a bound states. It is worth adding here that in the limit $m \to 1$, $K(m)$ tends to ∞ and the periodic nature of the potential is obscure. On the other hand, when $m = 0$, the Lamé potential (7.14) vanishes and one has a rigid rotator problem (of period $2K(m = 0) = \pi$), whose energy eigenvalues are at $E = 0, 1, 4, 9, \ldots$ with all the nonzero energy eigenvalues being two-fold degenerate.

Finally, it may be noted that the Schrödinger equation for finding the eigenstates for an arbitrary periodic potential is called Hill's equation in the mathematical literature. A general property of the Hill's equation is the oscillation theorem which states that for a potential with period L, the band edge wave functions arranged in order of increasing energy $E_0 \leq E_1 \leq E_2 \leq E_3 \leq E_4 \leq E_5 \leq E_6 \leq \dots$ are of period $L, 2L, 2L, L, L, 2L, 2L, \dots$. The corresponding number of (wave function) nodes in the interval L are $0, 1, 1, 2, 2, 3, 3, \dots$ and the energy band gaps are given by $\Delta_1 \equiv E_2 - E_1$, $\Delta_2 \equiv E_4 - E_3$, $\Delta_3 \equiv E_6 - E_5$, $\dots$. We shall see that the expected $m = 0$ limit and the oscillation theorem are very useful in making sure that all band edge eigenstates have been properly determined or if some have been missed.

Let us first consider the Lamé potential (7.14) with $a = 1$ and show that in this case the SUSY partner potentials are self-isospectral. The Schrödinger equation for the Lamé potential with $a = 1$ can be solved exactly and it is well known that in this case the spectrum consists of a single bound band and a continuum. In particular, the eigenstates for the lower and upper edge of the bound band are given by

$$E_0 = m; \quad \psi_0(x) = \operatorname{dn} x, \tag{7.21}$$

$$E_1 = 1; \quad \psi_1(x) = \operatorname{cn} x. \tag{7.22}$$

On the other hand, the eigenstate for the lower edge of the continuum band is given by

$$E_2 = 1 + m; \quad \psi_2(x) = \operatorname{sn} x. \tag{7.23}$$

Note that at $m = 0$ the energy eigenvalues are at $0, 1$ as expected for a rigid rotator and as $m \to 1$, one gets $V(x) \to 2 - 2 \operatorname{sech}^2 x$, the band width $1 - m$ vanishes as expected, and one has an energy level at $E = 0$ and the continuum begins at $E = 1$.

Using eq. (7.21) the corresponding superpotential turns out to be

$$W(x) = m \frac{\operatorname{sn} x \operatorname{cn} x}{\operatorname{dn} x}, \tag{7.24}$$

On making use of eq. (7.18) it is easily shown (see the problem at the end of the chapter) that this W satisfies the condition (7.8) and hence the corresponding partner potentials are indeed self-isospectral.

The Lamé potential (7.14) with $a = 1$, is one of the rare periodic potentials for which the dispersion relation between E and crystal momentum k is known in a closed form. This happens because the Schrödinger equation for this potential has two independent solutions given by

$$\psi(x) = \frac{H(x \pm \alpha)}{\Theta(x)} e^{\mp x Z(\alpha)}, \qquad (7.25)$$

where the parameter α is related to the energy eigenvalue E by $E = \mathrm{dn}^2(\alpha, m)$, $H(x)$ is the Jacobi eta function, $\Theta(x)$ is the Jacobi theta function, and $Z(\alpha)$ is the Jacobi zeta function. Using this exact solution and the Bloch condition, (7.1) one can find the dispersion relation by noting that the Lame potential is of period $2K(m)$. In particular, it can be shown that the dispersion relation is given by

$$k = \mp \frac{\pi}{2K(m)} \pm iZ(\mathrm{dn}^{-1}(\frac{\sqrt{E}}{m})). \qquad (7.26)$$

In Fig. 7.1 we have plotted this dispersion relation for the case $m = 0.3$ which clearly shows the band gap.

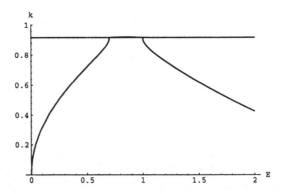

Fig. 7.1 The *exact* dispersion relation (7.26) between energy E and crystal momentum k for the Lamé potential (7.14) with $a = 1$. This plot is for $m = 0.3$. The horizontal line marks the edge of the Brillioun zone, at which $k = \frac{\pi}{2K(m)}$.

In view of this result for $a = 1$, one might think that even for higher integer values of a, the two partner potentials would be self-isospectral. However, this is not so, and in fact for any integer $a(\geq 2)$ we obtain a new exactly solvable periodic potential. As an illustration, consider the Lamé potential (7.14) with $a = 2$. For the $a = 2$ case, the Lamé potential has 2

Table 7.1 The eigenvalues and eigenfunctions for the 5 band edges corresponding to the $a = 2$ Lamé potential V_1 which gives $(p, q) = (6, 0)$ and its SUSY partner V_2. Here $B \equiv 1 + m + \delta$ and $\delta \equiv \sqrt{1 - m + m^2}$. The potentials $V_{1,2}$ have period $L = 2K(m)$ and their analytic forms are given by eqs. (7.27) and (7.30) respectively. The periods T of various eigenfunctions and the number of nodes N in the interval L are tabulated

E	$\psi^{(1)}$	$\psi_0^{(1)}\psi^{(2)}$	T	N
0	$B - 3m\,\mathrm{sn}^2 x$	1	$2K$	0
$3\delta - B$	$\mathrm{cn}\,x\,\mathrm{dn}\,x$	$\mathrm{sn}\,x[6m - (m+1)B$	$4K$	1
		$+ m\,\mathrm{sn}^2 x(2B - 3 - 3m)]$		
$2B - 3$	$\mathrm{sn}\,x\,\mathrm{dn}\,x$	$\mathrm{cn}\,x[B + m(3 - 2B)\mathrm{sn}^2 x]$	$4K$	1
$2B - 3m$	$\mathrm{sn}\,x\,\mathrm{cn}\,x$	$\mathrm{dn}\,x[B + (3m - 2B)\mathrm{sn}^2 x]$	$2K$	2
4δ	$B - 2\delta - 3m\,\mathrm{sn}^2 x$	$\mathrm{sn}\,x\,\mathrm{cn}\,x\,\mathrm{dn}\,x$	$2K$	2

bound bands and a continuum band. The energies and wave functions of the five band edges are well known. The lowest energy band ranges from $2 + 2m - 2\delta$ to $1 + m$, the second energy band ranges from $1 + 4m$ to $4 + m$ and the continuum starts at energy $2 + 2m + 2\delta$, where $\delta = \sqrt{1 - m + m^2}$. The wave functions of all the band edges are given in Table 7.1

Note that in the interval $2K(m)$ corresponding to the period of the Lamé potential, the number of nodes increases with energy. In order to use the SUSY QM formalism, we must shift the Lamé potential by a constant to ensure that the ground state (i.e. the lower edge of the lowest band) has energy $E = 0$. As a result, the potential

$$V_1(x) = -2 - 2m + 2\delta + 6m\,\mathrm{sn}^2 x, \qquad (7.27)$$

has its ground state energy at zero with the corresponding un-normalized wave function

$$\psi_0^{(1)}(x) = 1 + m + \delta - 3m\,\mathrm{sn}^2 x. \qquad (7.28)$$

The corresponding superpotential is

$$W = -\frac{d}{dx}\log\psi_0^{(1)}(x) = \frac{6m\mathrm{sn}\,x\ \ \mathrm{cn}\,x\,\mathrm{dn}\,x}{\psi_0^{(1)}(x)}, \qquad (7.29)$$

and hence the partner potential corresponding to (7.27) is

$$V_2(x) = -V_1(x) + \frac{72m^2\,\mathrm{sn}^2 x\,\mathrm{cn}^2 x\,\mathrm{dn}^2 x}{[1 + m + \delta - 3m\,\mathrm{sn}^2 x]^2}. \qquad (7.30)$$

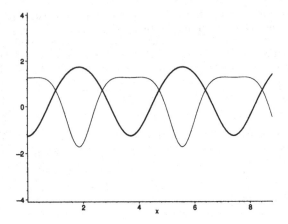

Fig. 7.2   The (6,0) Lamé potential $V_1(x)$ corresponding to $a = 2$ [thick line] as given by eq. (7.27) and its supersymmetric partner potential $V_2(x)$ [thin line] as given by eq. (7.30) for $m = 0.5$.

Although the SUSY QM formalism guarantees that the potentials $V_{1,2}$ are isospectral, they are not self-isospectral, since they do not satisfy eq. (7.9). Therefore, $V_2(x)$ as given by eq. (7.30) is a new periodic potential which is strictly isospectral to the potential (7.27) and hence it also has 2 bound bands and a continuum band. In Figs. 7.2 and 7.3 we have plotted the potentials $V_{1,2}(x)$ corresponding to $a = 2$ for two different values of the parameter m.

The difference in shape between $V_1(x)$ and $V_2(x)$ is manifest from the figures, especially for large m. Using eqs. (7.6) and (7.7) and the known eigenstates of $V_1(x)$, we can immediately compute all the band-edge Bloch wave functions for $V_2(x)$. In Table 7.1 we have given the energy eigenvalues and wave functions for the isospectral partner potentials $V_{1,2}(x)$. At $m = 0$ one has energy eigenvalues $0, 1, 4$ as expected for a rigid rotator. As $m \to 1$, one gets $V_1(x) \to 4 - 6 \operatorname{sech}^2 x$, the band widths vanish as expected, and one has two energy levels at $E = 0, 3$, with a continuum beginning from $E = 4$.

Finally, consider the $a = 3$ Lamé potential as given by eq. (7.14). The ground state wave function is known to be

$$\psi_0^{(1)}(x) = \operatorname{dn} x [2m + \delta_1 + 1 - 5m \operatorname{sn}^2 x] , \qquad (7.31)$$

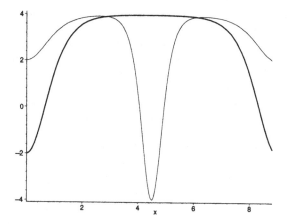

Fig. 7.3 Same parameters as Fig. 7.2 except for $m = 0.998$.

hence the corresponding superpotential is

$$W = \frac{m\,\mathrm{sn}\,x\,\mathrm{cn}\,x}{\mathrm{dn}\,x}\,\frac{[2m + \delta_1 + 11 - 15m\,\mathrm{sn}^2 x]}{[2m + \delta_1 + 1 - 5m\,\mathrm{sn}^2 x]}, \qquad (7.32)$$

and the partner potentials $V_{1,2}(x)$ are

$$V_1(x) = -2 - 5m + 2\delta_1 + 12m\,\mathrm{sn}^2 x, \quad \delta_1 \equiv \sqrt{1 - m + 4m^2}, \qquad (7.33)$$

and

$$V_2(x) = -V_1(x) + \frac{2m^2\mathrm{sn}^2 x\,\mathrm{cn}^2 x}{\mathrm{dn}^2 x}\,\frac{[2m + \delta_1 + 11 - 15m\,\mathrm{sn}^2 x]^2}{[2m + \delta_1 + 1 - 5m\,\mathrm{sn}^2 x]^2}. \qquad (7.34)$$

Clearly, the potentials $V_{1,2}(x)$ are not self-isospectral. In fact, $V_1(x)$ and $V_2(x)$ are distinctly different periodic potentials which have the same seven band edges corresponding to three bound bands and a continuum band. In Fig. 7.4 we have plotted the potentials $V_{1,2}(x)$ corresponding to $a = 3$ for the value $m = 0.5$.

It is clear from the figure that the potentials $V_2(x)$ and $V_1(x)$ have different shapes and are far from being self-isospectral. Using eqs. (7.6) and (7.7) and the known eigenstates of $V_1(x)$, we can immediately compute all the 7 band edges of $V_2(x)$ corresponding to the known 3 bound bands

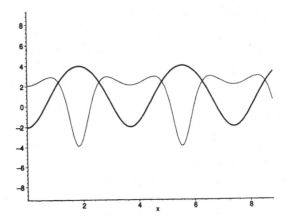

Fig. 7.4   The (12,0) Lamé potential $V_1(x)$ corresponding to $a = 3$ [thick line] as given by eq. (7.33) and its supersymmetric partner potential $V_2(x)$ [thin line] as given by eq. (7.34) for $m = 0.5$.

and a continuum band. For example, the ground state $\psi_0^{(2)}$ is given by

$$\psi_0^{(2)}(x) = \frac{1}{\psi_0^{(1)}(x)} = \frac{1}{\operatorname{dn} x[1 + 2m + \delta_1 - 5m \operatorname{sn}^2 x]} \,. \qquad (7.35)$$

The wave functions for the remaining six states are similarly written down by using eq. (7.7). These are shown in Table 7.2

Note that at $m = 0$ one has energy eigenvalues at $0, 1, 4, 9$ as expected for a rigid rotator and as $m \to 1$, one gets $V_1(x) \to 9 - 12 \operatorname{sech}^2 x$, the band widths vanish as expected, and one has three energy levels at $E = 0, 5, 8$ with a continuum above $E = 9$.

The extension to higher values of a is straightforward. It is possible to make several general comments about the form of the band edge wave functions for the partner potentials $V_2(x)$. This is most conveniently done by separately discussing the cases of even and odd values of a.

If a is an even integer, say $a = 2N$, it can be shown that the corresponding Lamé potential has $N + 1$ solutions of the form $F_N(\operatorname{sn}^2 x)$, and N solutions each of the three forms $\operatorname{sn} x \operatorname{cn} x F_{N-1}(\operatorname{sn}^2 x)$, $\operatorname{sn} x \operatorname{dn} x F_{N-1}(\operatorname{sn}^2 x)$, $\operatorname{cn} x \operatorname{dn} x F_{N-1}(\operatorname{sn}^2 x)$. Here F_r denotes a polynomial of degree r in its argument. The ground state $\psi_0^{(1)}(x)$ is of the form $F_N(\operatorname{sn}^2 x)$. It is easily checked using eq. (7.7) that the corresponding partner potential $V_2(x)$ has

Table 7.2 The eigenvalues and eigenfunctions for the 7 band edges corresponding to the $a = 3$ Lamé potential V_1 which gives $(p,q) = (12,0)$ and its SUSY partner V_2. Here $\delta_1 \equiv \sqrt{1 - m + 4m^2}$; $\delta_2 \equiv \sqrt{4 - m + m^2}$; $\delta_3 \equiv \sqrt{4 - 7m + 4m^2}$. The potentials $V_{1,2}$ have period $L = 2K(m)$ and their analytic forms are given by eqs. (7.33) and (7.34) respectively. The periods T of various eigenfunctions and the number of nodes N in the interval L are tabulated.

E	$\psi^{(1)}$	$\psi_0^{(1)}\psi^{(2)}$	T	N
0	$\mathrm{dn}x[1 + 2m + \delta_1 - 5m\mathrm{sn}^2x]$	1	$2K$	0
$3 - 3m + 2\delta_1 - 2\delta_2$	$\mathrm{cn}\,x[2 + m + \delta_2 - 5m\mathrm{sn}^2x]$	$10m(1 - m + \delta_2 - \delta_1)\times \mathrm{sn}\,x\,\mathrm{cn}^2x\,\mathrm{dn}^2x$ $-(1 - m)\frac{\mathrm{sn}x\,\psi_0^{(1)}\psi^{(1)}}{\mathrm{cn}x\,\mathrm{dn}x}$	$4K$	1
$3 + 2\delta_1 - 2\delta_3$	$\mathrm{sn}\,x[2 + 2m + \delta_3 - 5m\mathrm{sn}^2x]$	$10m(1 + \delta_3 - \delta_1)\mathrm{cn}\,x\,\mathrm{sn}^2x\,\mathrm{dn}^2x$ $-(1 - 2m\mathrm{sn}^2x)\frac{\mathrm{cn}x\,\psi_0^{(1)}\psi^{(1)}}{\mathrm{sn}x\,\mathrm{dn}x}$	$4K$	1
$2 - m + 2\delta_1$	$\mathrm{sn}\,x\,\mathrm{cn}x\,\mathrm{dn}\,x$	$\mathrm{dn}^3x[1 + 2m + \delta_1$ $+(m - 2 - 2\delta_1)\mathrm{sn}^2x]$	$2K$	2
$4\delta_1$	$\mathrm{dn}x[1 + 2m - \delta_1 - 5m\mathrm{sn}^2x]$	$\mathrm{sn}\,x\mathrm{cn}x\mathrm{dn}^3x$	$2K$	2
$3 - 3m + 2\delta_1 + 2\delta_2$	$\mathrm{cn}\,x[2 + m - \delta_2 - 5m\mathrm{sn}^2x]$	$10m(1 - m - \delta_2 - \delta_1)\times \mathrm{sn}\,x\,\mathrm{cn}^2x\,\mathrm{dn}^2x$ $-(1 - m)\frac{\mathrm{sn}x\,\psi_0^{(1)}\psi^{(1)}}{\mathrm{cn}x\,\mathrm{dn}x}$	$4K$	3
$3 + 2\delta_1 + 2\delta_3$	$\mathrm{sn}\,x[2 + 2m - \delta_3 - 5m\mathrm{sn}^2x]$	$10m(1 - \delta_3 - \delta_1)\mathrm{cn}\,x\,\mathrm{sn}^2x\,\mathrm{dn}^2x$ $-(1 - 2m\mathrm{sn}^2x)\frac{\mathrm{cn}x\,\psi_0^{(1)}\psi^{(1)}}{\mathrm{sn}x\,\mathrm{dn}x}$	$4K$	3

N solutions each of the four forms

$$\frac{\mathrm{dn}x\,G_N(\mathrm{sn}^2x)}{\psi_0^{(1)}(x)}\,, \quad \frac{\mathrm{sn}x\,G_N(\mathrm{sn}^2x)}{\psi_0^{(1)}(x)}\,,$$

$$\frac{\mathrm{cn}x\,G_N(\mathrm{sn}^2x)}{\psi_0^{(1)}(x)}\,, \quad \frac{\mathrm{sn}x\,\mathrm{cn}x\,\mathrm{dn}x\,G_{N-1}(\mathrm{sn}^2x)}{\psi_0^{(1)}(x)}\,,$$

while the ground state is given by $\psi_0^{(2)}(x) = 1/\psi_0^{(1)}(x)$.

If a is an odd integer, say $a = 2N + 1$, the corresponding Lamé potential has $N + 1$ solutions each of the three forms $\mathrm{sn}x\,F_N(\mathrm{sn}^2x)$, $\mathrm{cn}x\,F_N(\mathrm{sn}^2x)$, $\mathrm{dn}x\,F_N(\mathrm{sn}^2x)$, and N solutions of the form $\mathrm{sn}x\,\mathrm{cn}x\,\mathrm{dn}x\,F_{N-1}(\mathrm{sn}^2x)$. The ground state $\psi_0^{(1)}(x)$ is of the form $\mathrm{dn}x\,F_N(\mathrm{sn}^2x)$. We can then easily deduce that the corresponding partner potentials $V_2(x)$ will have $N + 1$ solutions each of the two forms

$$\frac{\mathrm{sn}x\,G_{N+1}(\mathrm{sn}^2x)}{\psi_0^{(1)}(x)}\,, \quad \frac{\mathrm{cn}x\,G_{N+1}(\mathrm{sn}^2x)}{\psi_0^{(1)}(x)}\,,$$

and N solutions each of the two forms

$$\frac{\mathrm{dn}x \; G_{N+1}(\mathrm{sn}^2 x)}{\psi_0^{(1)}(x)} \; , \; \frac{\mathrm{sn}x \; \mathrm{cn}x \; \mathrm{dn}x \; G_N(\mathrm{sn}^2 x)}{\psi_0^{(1)}(x)} \; ,$$

while as usual, the ground state is given by $\psi_0^{(2)}(x) = 1/\psi_0^{(1)}(x)$.

In summary, for integral a, Lamé potentials with $a \geq 2$ are not self isospectral. They have distinct supersymmetric partner potentials even though both potentials have the same $(2a + 1)$ band edge eigenvalues.

7.3 Associated Lamé Potentials and Their Supersymmetric Partners

We shall now discuss a much richer class of periodic potentials given by

$$V(x) = pm \, \mathrm{sn}^2 x + qm \frac{\mathrm{cn}^2 x}{\mathrm{dn}^2 x} \; ; \quad p \equiv a(a + 1) \, ; q \equiv b(b + 1) \; . \quad (7.36)$$

The potentials of eq. (7.36) are called associated Lamé potentials, since the corresponding Schrödinger equation is called the associated Lamé equation. We shall often refer to this potential as the (p, q) potential. Note that the (q, p) potential is just the (p, q) potential shifted by $K(m)$. Also the $(p, 0)$ potential is just the Lamé potential of eq. (7.14).

In general, for any value of p and q, the associated Lamé potentials have a period $2K(m)$ since

$$\mathrm{sn}(x + 2K) = -\mathrm{sn} \, x \; , \; \mathrm{cn}(x + 2K) = -\mathrm{cn} \, x \; , \; \mathrm{dn}(x + 2K) = \mathrm{dn} \, x \; .$$

However, for the special case $p = q$, eq. (7.18) shows that the period is $K(m)$. From a physical viewpoint, if one thinks of a Lamé potential $(p, 0)$ as due to a one-dimensional regular array of atoms with spacing $2K(m)$, and "strength" p, then the associated Lamé potential (p, q) results from two alternating types of atoms spaced by $K(m)$ with "strengths" p and q respectively. If the two types of atoms are identical (which makes $p = q$), one expects a potential of period $K(m)$.

We start with the associated Lamé equation which is just the Schrödinger equation for the potential (7.36)

$$-\frac{d^2\psi}{dx^2} + [pm \, \mathrm{sn}^2 x + qm \frac{\mathrm{cn}^2 x}{\mathrm{dn}^2 x} - E]\psi = 0 \; . \quad (7.37)$$

On substituting

$$\psi(x) = [\mathrm{dn}\, x]^{-b}\, y(x)\,, \tag{7.38}$$

it is easily shown that $y(x)$, satisfies the Hermite elliptic equation

$$y''(x) + 2bm\frac{\mathrm{sn}\, x\, \mathrm{cn}\, x}{\mathrm{dn}\, x} y'(x) + [\lambda - (a+1-b)(a+b)m\, \mathrm{sn}^2 x] y(x) = 0\,, \tag{7.39}$$

where

$$p = a(a+1)\,, \quad q = b(b+1)\,, \quad E = \lambda + mb^2\,. \tag{7.40}$$

On further substituting

$$\mathrm{sn}\, x = \sin t\,, \quad y(x) \equiv z(t)\,, \tag{7.41}$$

one obtains Ince's equation

$$(1 - m\sin^2 t)z''(t) + (2b-1)m\sin t\cos t\, z'(t)$$
$$+[\lambda - (a+1-b)(a+b)m\sin^2 t]z(t) = 0\,, \tag{7.42}$$

which is a well known quasi-exactly solvable (QES) equation. On substituting

$$\cos t = u\,, \quad z(t) \equiv w(u) = \sum_{n=0}^{\infty} \frac{u^n R_n}{n!}\,, \tag{7.43}$$

it is easily shown that R_n satisfies a three-term recursion relation. In particular if $a + b + 1 = n$ ($n = 1, 2, 3, ...$) then one obtains n QES solutions. Actually n QES solutions are also obtained for $b - a = -n (n = 1, 2, 3, ...)$ but since q is unchanged under $b \to -b - 1$, no really new solutions are obtained in this case. The QES solutions for $n = 1, 2, 3, 4, 5$ are given in Table 7.3. In particular, for any given choice of $p = a(a+1)$, Table 7.3 lists the eigenstates of the associated Lamé equation for various values of q.

It is easily checked from Table 7.3 that the solution corresponding to $q = a(a-1)$ as well as one of the $q = (a-2)(a-3)$ solutions are nodeless and correspond to the ground state. Hence, for these cases, one can obtain the superpotential and hence the partner potential V_2 and enquire if $V_{1,2}$ are self-isospectral or not.

Let us now consider the SUSY partner potentials computed from the ground state for the $p = a(a+1), q = (a-3)(a-2)$ case. It is given by

Table 7.3   Eigenvalues and eigenfunctions for various associated Lamé potentials (p,q) with $p = a(a+1)$ and $q = (a-n+1)(a-n)$ for $n = 1,2,3,\ldots$. The periods of various eigenfunctions and the number of nodes in the interval $2K(m)$ are tabulated. Here $\delta_4 \equiv \sqrt{1-m+m^2(a-1)^2}$; $\delta_5 \equiv \sqrt{4-7m+2ma+m^2(a-2)^2}$; $\delta_6 \equiv \sqrt{4-m-2ma+m^2(a-1)^2}$; $\delta_7 \equiv \sqrt{9-9m+m^2(a-2)^2}$.

q	E	$dn^{-a}(x)\psi$	T	N
$a(a-1)$	ma^2	1	$2K$	0
$(a-1)(a-2)$	$1+m(a-1)^2$	$\frac{cn\,x}{dn\,x}$	$4K$	1
$(a-1)(a-2)$	$1+ma^2$	$\frac{sn\,x}{dn\,x}$	$4K$	1
$(a-2)(a-3)$	$m(a^2-2a+2)$ $+2\pm 2\delta_4$	$\frac{1}{dn^2x}[m(2a-1)sn^2x$ $-1+m(1-a)\pm\delta_4]$	$2K$	2,0
$(a-2)(a-3)$	$4+m(a-1)^2$	$\frac{sn\,x\,cn\,x}{dn^2x}$	$2K$	2
$(a-3)(a-4)$	$m(a^2-4a+5)$ $+5\pm 2\delta_5$	$\frac{cn\,x}{dn^3x}[m(2a-1)sn^2x$ $-2+m(2-a)\pm\delta_5]$	$4K$	3,1
$(a-3)(a-4)$	$m(a^2-2a+2)$ $+5\pm 2\delta_6$	$\frac{sn\,x}{dn^3x}[m(2a-1)sn^2x$ $-2+m(1-a)\pm\delta_6]$	$4K$	3,1
$(a-4)(a-5)$	$m(a^2-4a+5)$ $+10\pm 2\delta_7$	$\frac{sn\,x\,cn\,x}{dn^4x}[m(2a-1)sn^2x$ $-3+m(2-a)\pm\delta_7]$	$2K$	4,2

(see Table 7.3)

$$\psi_0(x) = \left[m(1-a) - 1 - \delta_4 + m(2a-1)sn^2x \right](dn\,x)^{a-2} , \qquad (7.44)$$

where $\delta_4 = \sqrt{1-m+m^2(a-1)^2}$. The corresponding superpotential W turns out to be

$$W = \frac{m(a-2)sn\,x\,cn\,x}{dn\,x} - \frac{2m(2a-1)sn\,x\,cn\,x\,dn\,x}{[m(1-a) - 1 - \delta_4 + m(2a-1)sn^2x]} . \qquad (7.45)$$

Hence the corresponding partner potentials are

$$V_1(x) = ma(a+1)sn^2x$$
$$+ m(a-3)(a-2)\frac{cn^2x}{dn^2x} - 2 - m(a^2-2a+2) + 2\delta_4, \qquad (7.46)$$

$$V_2(x) = -V_1(x) + 2W^2(x) . \qquad (7.47)$$

It is easily checked that these potentials are not self-isospectral since they do not satisfy the condition (7.9). Thus one has discovered a whole class of new elliptic periodic potentials $V_2(x)$ as given by eq. (7.47) for which three states are analytically known no matter what a is. In particular, the energy eigenfunctions for V_2 of these three states are easily obtained by using the corresponding energy eigenstates of V_1 as given in Table 7.3 and using eqs. (7.6) and (7.7).

We might add here that the well known solutions for the Lamé potential (7.14) with integer a are contained in Table 7.3. For example, when $a = 3$, one has the (12,0) potential. From Table 7.3 it follows that the 3 band edges of period $2K(m)$ are obtained from $q = (a - 2)(a - 3)$ and 4 band edges of period $4K(m)$ are obtained from $q = (a - 3)(a - 4)$. Altogether, arranging in order of increasing nodes, one has 7 band edges with periods $2K, 4K, 4K, 2K, 2K, 4K, 4K$ with $0, 1, 1, 2, 2, 3, 3$ nodes respectively. There are no missing states and this gives three bound bands and a continuum band.

From Table 7.3 it is also clear that if a and b are either both integers or half-integers then several band edge energies are exactly known though in most cases one usually does not know all the band edge energies, that is one has a QES problem. However, in the special case of $p = q$=integer ($a = b =$ integer), we show that all the band edge eigenstates can be obtained and one has an exactly solvable periodic problem.

7.3.1 $a = b = $ *Integer*

Let us now discuss the special case of $p = q = a(a+1)$, $a = 1, 2, ...$. In this case the associated Lamé potential (7.36) has period K, rather than $2K$. It then follows from the oscillation theorem that with increasing energy, the band edges must have periods $K, 2K, 2K, K, K, ...$ and in the $m = 0$ limit the eigenvalues must go to $E = 0, 4, 16, 36, ...$ with all nonzero eigenvalues being doubly degenerate. One case for which we already have exact results is when $p = q = 2$. In particular, using eqs. (7.44) to (7.47) and taking $a = 1$ in $q = (a - 2)(a - 3)$ we can calculate three energy eigenvalues and eigenfunctions of V_1 (see Problem 7.3 at the end of the chapter) given by

$$V_1(x) = 2m\,\text{sn}^2 x + 2m\frac{\text{cn}^2 x}{\text{dn}^2 x} - 2 - m + 2\sqrt{1 - m}\,. \qquad (7.48)$$

Whereas the ground state is of period K, the next two states in Table 7.3 indeed have period $2K$. Using $a = 1$ in eqs. (7.44) to (7.47), we find that the corresponding SUSY partner potential is

$$V_2(x) = 2 - m - 2\sqrt{1-m} - \frac{8\sqrt{1-m}\, m^2 \mathrm{sn}^2 x\, \mathrm{cn}^2 x}{[\mathrm{dn}^2 x + \sqrt{1-m}]^2}. \tag{7.49}$$

Are the potentials $V_{1,2}(x)$ self-isospectral? Using the relations

$$\mathrm{sn}(x + K(m)/2) = (1 + \sqrt{1-m})^{\frac{1}{2}} \left[\frac{\sqrt{1-m}\,\mathrm{sn}\,x + \mathrm{cn}\,x\,\mathrm{dn}\,x}{\mathrm{dn}^2 x + \sqrt{1-m}} \right], \tag{7.50}$$

$$\mathrm{cn}(x + K(m)/2)$$
$$= (1 + \sqrt{1-m})^{1/2}(1-m)^{\frac{1}{4}} \left[\frac{(1+\sqrt{1-m})^{1/2}\mathrm{cn}\,x - \mathrm{sn}\,x\,\mathrm{dn}\,x}{\mathrm{dn}^2 x + \sqrt{1-m}} \right], \tag{7.51}$$

$$\mathrm{dn}(x + K(m)/2) \quad = (1-m)^{\frac{1}{4}} \left[\frac{(1+\sqrt{1-m})\mathrm{dn}\,x - m\mathrm{sn}\,x\,\mathrm{cn}\,x}{\mathrm{dn}^2 x + \sqrt{1-m}} \right], \tag{7.52}$$

a little algebra reveals that indeed $V_{1,2}$ are self-isospectral and satisfy eq. (7.9).

Are the higher members of the $p = q$ family (i.e. $p = q = 6, 12, 20, ...$) also self-isospectral? If our experience with the Lamé case is any guide then we would doubt it. Indeed, we will now show that the (6,6) associated Lamé potential is not self-isospectral. First of all let us note that for this case we get five band edges analytically from Table 7.3. In particular, take $a = 2$ and consider the case of $q = (a-4)(a-5)$, for which we know two eigenstates as given in Table 7.3. In fact, in this case three more eigenstates can be analytically obtained but the corresponding eigenvalues and eigenfunctions have not been given in Table 7.3 since the energy eigenvalues are solutions of a cubic equation whose exact solution for arbitrary a can not be written in a compact form. However, for $a = 2$, we are able to solve the cubic equation and obtain the three eigenvalues in a closed simple form. In particular consider an ansatz of the form

$$y = A + B\mathrm{sn}^2 x + D\mathrm{sn}^4 x. \tag{7.53}$$

On substituting this ansatz in eq. (7.39) it is easy to show that the energy

Table 7.4 The five eigenvalues and eigenfunctions for the associated Lamé potential corresponding to $a = b = 2$ which gives $(p, q) = (6, 6)$. Here $\delta_8 \equiv \sqrt{16 - 16m + m^2}$. The number of nodes in one period $K(m)$ of the potential is tabulated.

E	$\mathrm{dn}^2 x \psi^{(1)}$	Period	Nodes
0	$1 - (4 - m - \delta_8)\mathrm{sn}^2 x$ $+ (4 - 2m - \delta_8)\mathrm{sn}^4 x$	K	0
$-4 + 2m + 2\delta_8$	$1 - 2\mathrm{sn}^2 x + m\,\mathrm{sn}^4 x$	$2K$	1
$2 - m + 2\delta_8$ $-6\sqrt{1 - m}$	$\mathrm{sn}\,x\,\mathrm{cn}\,x[1$ $-(1 - \sqrt{1 - m})\mathrm{sn}^2 x]$	$2K$	1
$2 - m + 2\delta_8$ $+6\sqrt{1 - m}$	$\mathrm{sn}\,x\,\mathrm{cn}\,x[1$ $-(1 + \sqrt{1 - m})\mathrm{sn}^2 x]$	K	2
$4\delta_8$	$1 - (4 - m + \delta_8)\mathrm{sn}^2 x$ $+ (4 - 2m + \delta_8)\mathrm{sn}^4 x$	K	2

eigenvalue $\lambda(= E - m(a - 4)^2)$ must obey the cubic equation

$$\lambda^3 + 4[7m - 5 - 3am]\lambda^2 + 16[4 + m(10a - 19)$$
$$+ 2m^2(a - 2)(a - 3)]\lambda - 64m(2a - 3)(2 - 2m + ma) = 0. \quad (7.54)$$

The solution of this equation is in general quite lengthy but in the special case of $a = 2$ this cubic equation is easily solved yielding three eigenvalues in a compact form. On combining them with the two levels given in Table 7.3, we obtain the eigenvalues and eigenfunctions of all the five band edges for the case $p = q = 6$. These are given in Table 7.4.

We have also verified that these five eigenstates in ascending order of energy indeed have periods $K, 2K, 2K, K, K$ respectively and that the energy eigenvalues have expected limits at $m = 0$. In particular the associated Lamé potential $V_1(x)$ is

$$V_1(x) = 6m\,\mathrm{sn}^2 x + 6m\frac{\mathrm{cn}^2 x}{\mathrm{dn}^2 x} - 8 - 2m + 2\delta_8\,, \quad (7.55)$$

whose ground state energy is zero while the corresponding eigenfunction $\psi_0^{(1)}$ is

$$\psi_0^{(1)}(x) = \frac{\left[1 - (4 - m - \delta_8)\mathrm{sn}^2 x + (4 - 2m - \delta_8)\mathrm{sn}^4(x)\right]}{\mathrm{dn}^2 x}\,, \quad (7.56)$$

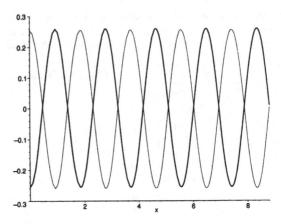

Fig. 7.5   The (6,6) associated Lamé potential $V_1(x)$ [thick line] as given by eq. (7.55) and its supersymmetric partner potential $V_2(x)$ [thin line] as given by eq. (7.58) for $m = .5$.

where $\delta_8 = \sqrt{16 - 16m + m^2}$. Hence the corresponding superpotential is

$$W(x) = \frac{-2m\operatorname{sn}x\operatorname{cn}x}{\operatorname{dn}x} + \frac{2\operatorname{sn}x\operatorname{cn}x}{\operatorname{dn}x\,\psi_0^{(1)}(x)}\left[(4 - m - \delta_8) - 2(4 - 2m - \delta_8)\operatorname{sn}^2x\right],$$

(7.57)

and the partner potential $V_2(x)$ which is isospectral to $V_1(x)$ is

$$V_2(x) = -V_1(x) + 2W^2(x). \tag{7.58}$$

It is easily shown that $W(x)$ as given by eq. (7.57) does not satisfy the self-isospectrality condition (7.8). Hence, unlike the $p = q = 2$ case, the $p = q = 6$ potential is *not* self-isospectral. In Figs. 7.5 and 7.6 we have plotted the potentials $V_{1,2}(x)$ corresponding to $p = q = 6$ for two values of the parameter m. The figures confirm that the potentials are far from being self-isospectral. Thus we have obtained a new exactly solvable periodic potential (7.58) which has two bound bands and a continuum band, with five band edges and the corresponding eigenfunctions being exactly known using Table 7.4 and eqs. (7.6) and (7.7).

It is clear that the higher associated Lamé potentials with $p = q = 12, 20, \ldots$ which have $7, 9, \ldots$ band edges are also exactly solvable in principle and none of them will be self-isospectral, so that in each case one obtains a new exactly solvable periodic potential. In particular, for $p = q = n(n + 1)$ there will be $(2n + 1)$ band edges in both $V_{1,2}(x)$ whose energy eigenvalues

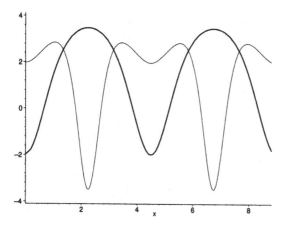

Fig. 7.6 Same as Fig. 7.5 but $m = .998$.

can be obtained from Table 7.3 when q has the form $[n - 2n][n - (2n + 1)]$. Out of the $(2n + 1)$ band edges in $V_1(x)$, $(n + 1)$ solutions (including the ground state) have the form $\frac{F_n(\text{sn}^2 x)}{\text{dn}^n x}$ while n solutions have the form $F_{n-1}(\text{sn}^2 x) \frac{\text{sn}\,x\,\text{cn}\,x}{\text{dn}^n x}$.

On the other hand, as far as the $(2n + 1)$ solutions of the partner potential V_2 are concerned, there are n states each of the two forms

$$\frac{\text{sn}\,x\,\text{cn}\,x\,G_n(\text{sn}^2 x)}{\text{dn}^{2n-1} x\,\psi_0^{(1)}(x)} , \frac{G_{n+1}(\text{sn}^2 x)}{\text{dn}^{2n-1} x\,\psi_0^{(1)}(x)} ,$$

while the ground state is given by $\psi_0^{(2)}(x) = 1/\psi_0^{(1)}(x)$.

Thus, using the formalism of SUSY QM, one is able to discover many new exactly solvable and quasi-exactly periodic potentials involving Jacobi elliptic functions.

References

(1) For the properties of Jacobi elliptic functions, see, for example, I. S. Gradshteyn and I. M. Ryzhik, *Table of Integrals, Series and Products*, Academic Press, (1980). The modulus parameter m is often called k^2 in the mathematics literature.

(2) F.M. Arscott, *Periodic Differential Equations*, Pergamon(1981);

E.T. Whittaker and G.N. Watson, *A Course of Modern Analysis*, Cambridge University Press (1980).

(3) W. Magnus and S. Winkler, *Hill's Equation*, Wiley (1966).

(4) See, for example, F. Cooper, A. Khare and U. P. Sukhatme, *Supersymmetry and Quantum Mechanics*, Phys. Rep. **251** (1995) 267-385.

(5) G. Dunne and J. Feinberg, *Self Isospectral Periodic Potentials and Supersymmetric Quantum Mechanics*, Phys. Rev. **D57** (1998) 1271-1276.

(6) G. Dunne and J. Mannix, *Supersymmetry Breaking with Periodic Potentials*, Phys. Lett. **B428** (1998) 115-119.

(7) A. Khare and U. Sukhatme, *New Solvable and Quasi Exactly Solvable Periodic Potentials* , Jour. Math. Phys. **40** (1999) 5473-5494.

(8) A. Turbiner, *Quasi Exactly Solvable Problems and SL(2) Group*, Comm. Math. Phys. **118** (1988) 467-474.

(9) A.G. Ushveridze, *Quasi Exactly Solvable Models in Quantum Mechanics*, IOP Publishing Co. (1993).

Problems

1. Obtain the partner potentials corresponding to the periodic superpotential $W(x) = A \sin x$. Is SUSY broken or unbroken in this case? Are the partner potentials self-isospectral?

2. Check that the Lamé potential $V_1(x) = 2m\mathrm{sn}^2(x,m) - m$ has a zero energy eigenstate with eigenfunction $\psi_0(x) = \mathrm{dn}(x,m)$. Prove that this potential is self-isospectral by explicitly working out its SUSY partner potential.

3. Find the eigenvalues E_0, E_1, E_2, and the corresponding eigenfunctions for the associated Lamé potential with $a = b = 1$. Verify the result for the partner potential $V_2(x)$ given in the text and show that $V_1(x)$ and $V_2(x)$ are self-isospectral.

4. Using Table 7.3 determine the ground state of the associated Lamé potential corresponding to $(a,b) = (2,1)$. Work out the SUSY partner and check if it is self-isospectral.

Chapter 8

Supersymmetric WKB Approximation

WKB theory is a very successful method for obtaining global approximations to solutions of ordinary differential equations. It has numerous applications in physics and mathematics. Even though some general mathematical techniques were developed in the early nineteenth century, systematic development took place only after the emergence of quantum mechanics. WKB theory is applicable to differential equations when the highest derivative has a small multiplicative parameter ϵ. Such situations occur in boundary-value and Sturm-Liouville problems, and in particular in quantum mechanics, the small parameter ϵ is related to the quantity $\hbar^2$ in Schrödinger's equation. When applied to quantum mechanics, WKB theory is often called the semiclassical method since it enables one to take the parameter $\hbar$ to zero and study the classical limit. It has been successfully used for many years to determine eigenvalues and to compute barrier tunneling probabilities. The analytic properties of the WKB approximation have been studied in detail from a purely mathematical point of view, and the accuracy of the method has been tested by comparison between analytic and numerical results. An excellent review of WKB theory from a mathematical point of view is given in the book of Bender and Orszag.

In this chapter we will first review the main results of WKB theory. Then we will describe a recent extension of the semiclassical approach inspired by supersymmetry called the supersymmetric WKB (SWKB) method. We will show that for many problems the SWKB method gives better accuracy than the WKB method. In particular, we discuss and prove the remarkable result that the lowest order SWKB approximation gives exact energy eigenvalues for all simple SIPs with translation.

8.1 Lowest Order WKB Quantization Condition

The semiclassical WKB approximation for one dimensional potentials with two classical turning points is discussed in most quantum mechanics textbooks. Let us look at the standard situation of a potential on the entire real line, which has two classical turning points x_L and x_R given by $V(x) = E$ for any choice of energy $E(> V_{min})$. To derive the WKB quantization condition we have to connect the solution in the classically allowed region with the solution in the classically forbidden region.

We start from the time independent Schrödinger equation

$$-\frac{\hbar^2}{2m}\psi''(x) + [V(x) - E]\psi(x) = 0 \,. \tag{8.1}$$

In the WKB approximation, one substitutes

$$\psi(x) = A\, e^{\frac{iS(x)}{\hbar}} \,, \tag{8.2}$$

in the Schrödinger eq. (8.1). One then finds that $S(x)$ satisfies

$$(S')^2 - i\hbar S'' = 2m(E - V) \,. \tag{8.3}$$

One now substitutes an expansion of S in powers of $\hbar$

$$S = S_0 + \hbar S_1 + \dots \,, \tag{8.4}$$

in eq. (8.3) and equate equal powers of $\hbar$ to obtain a sequence of equations

$$
\begin{aligned}
(S_0')^2 &= 2m(E - V) \,, \\
iS_0'' &= 2S_0' S_1' \,, \\
&\;\;\vdots \,.
\end{aligned}
\tag{8.5}
$$

The first two equations give

$$
\begin{aligned}
S_0(x) &= \pm\hbar\chi(x) \,, \\
S_1(x) &= \frac{i}{2}\ln[p(x, E)] \,,
\end{aligned}
\tag{8.6}
$$

where arbitrary constants of integration that can be absorbed in A have been omitted. Here $p(x, E)$ is the generalized momentum defined by

$$p(x, E) = \sqrt{2m \mid E - V(x) \mid}, \tag{8.7}$$

while $\chi(x)$ is

$$\chi(x) = \frac{1}{\hbar} \int^x p(x, E) dx. \tag{8.8}$$

To this order of approximation, the solution, in the classically allowed region $x_L \leq x \leq x_R$, is given by

$$\psi_I^{(1)}(x) = \frac{A}{\sqrt{p(x, E)}} e^{\pm i\chi(x)}, \quad V < E, \tag{8.9}$$

while in the classically forbidden region $x > x_R$ or $x < x_L$ it is given by

$$\psi_{II,III}^{(1)}(x) = \frac{A}{\sqrt{p(x, E)}} e^{\pm \chi(x)}, \quad V > E. \tag{8.10}$$

At this point it is worth digressing a minute and examine the region of validity of the WKB approximation. Since S_0 is a monotonic increasing function of x (so long as p does not vanish), hence the ratio $\hbar S_1/S_0$ is expected to be small if $\hbar S_1'/S_0'$ is small. Hence the solution (8.9) is expected to be a useful solution so long as

$$\mid \frac{\hbar S_1'}{S_0'} \mid = \mid \frac{p'}{2p^2} \mid \ll 1. \tag{8.11}$$

Since the local de Broglie wavelength λ is $2\pi/p$, hence this condition can also be written as

$$\frac{\lambda}{4\pi} \mid \frac{dp}{dx} \mid \ll p. \tag{8.12}$$

We thus conclude that the WKB approximation is useful when the de Broglie wavelength λ is small compared to the characteristic distance over which the potential varies appreciably. In other words, the potential must be essentially constant over many wavelengths i.e. the WKB approximation is reliable in the short-wavelength limit.

We now observe that the condition (8.11) is badly violated near the turning points x_L, x_R for which $V(x) = E$. In fact it is a nontrivial task to match the two solutions (8.9) and (8.10) across the classical turning points x_L and x_R. This has been discussed in great detail in the literature so we shall merely quote the results of such an analysis. The standard procedure is to make a linear approximation to the potential near the turning points x_L, x_R and solve the resulting Schrödinger equation for a linear potential which is valid near the turning points x_L, x_R. One then has to match this solution to the other two as given by eqs. (8.9) and (8.10) by appropriately choosing various constants of integration. In this way one obtains the famous WKB quantization condition

$$\int_{x_L}^{x_R} dx \sqrt{2m[E - V(x)]} = (n + 1/2)\hbar\pi \,, \qquad (8.13)$$

where $n = 0, 1, 2, \ldots$. It is easily seen that here n denotes the number of nodes of the WKB wave function between the turning points.

It is worth recalling that the WKB wave functions (8.9) and (8.10) diverge at the classical turning points x_L, x_R. Although this divergence is understandable in the classical limit, since a classical particle has zero speed at the turning points, it is certainly not present in a full quantum mechanical treatment. It is because of this divergence that one has to resort to connection formulas and somewhat tricky matching of the wave functions that eventually yields the well known WKB quantization condition eq. (8.13).

8.1.1　*Simpler Approach for the Lowest Order Quantization Condition*

Very recently, a much simpler heuristic derivation of the WKB quantization condition (8.13) has been given. In this approach one effectively matches the zeroth order nondivergent wave function (coming solely from S_0) at the classical turning points. For simplicity, we restrict our attention to symmetric potentials $V(x) = V(-x)$. For this case, $x_L = -x_R$, and it is sufficient to just look at the half line $x > 0$, since the eigenfunctions will be necessarily symmetric or antisymmetric. Using eqs. (8.2) to (8.8) it follows that for the symmetric case, the zeroth order WKB wave function in the classically allowed region $x_L \leq x \leq x_R$ is

$$\psi_I^{(0)}(x) = A \cos[\chi(x) - \chi(0)] \,, \qquad (8.14)$$

while in the classically forbidden region II $(x > x_R)$ it is given by

$$\psi_{II}^{(0)}(x) = B e^{-\chi(x) + \chi(x_R)} \,. \qquad (8.15)$$

Matching the wave functions $\psi_I^{(0)}(x)$ and $\psi_{II}^{(0)}(x)$ and their first derivatives at x_R gives two equations

$$A\cos[\chi(x_R) - \chi(0)] = B, \tag{8.16}$$

$$A\sin[\chi(x_R) - \chi(0)] = B, \tag{8.17}$$

which yield $\tan[\chi(x_R) - \chi(0)] = 1$, or

$$\frac{1}{\hbar}\int_0^{x_R} p(x,E)dx = \frac{1}{4}\pi, \frac{5}{4}\pi, \frac{9}{4}\pi, \ldots. \tag{8.18}$$

Similarly for the antisymmetric case, the zeroth order WKB approximation to the wave function in the classically allowed region $x_L \leq x \leq x_R$ is $\psi_I^{(0)}(x) = A\sin[\chi(x) - \chi(0)]$ while in the classically forbidden region II ($x > x_R$) it is $\psi_{II}^{(0)}(x) = Be^{-\chi(x)+\chi(x_R)}$. Matching these wave functions and their first derivatives at x_R now gives $\tan[\chi(x_R) - \chi(0)] = -1$, or

$$\frac{1}{\hbar}\int_0^{x_R} p(x,E)dx = \frac{3}{4}\pi, \frac{7}{4}\pi, \frac{11}{4}\pi, \ldots. \tag{8.19}$$

Combining eqs. (8.18) and (8.19), we then obtain the quantization condition (8.13). This derivation is evidently much simpler than the usual textbook approach for deriving connection formulas.

Why is this simple procedure for matching $\psi^{(0)}(x)$ justified? Clearly, the correct approach is neither to match $\psi^{(0)}(x)$ nor $\psi^{(1)}(x)$, but to keep a sufficient number of higher order contributions in $\hbar$, so that the resulting wave function is non-divergent. This has to be the case, since there is no divergence in the full wave function. A simple way in which the divergence gets tamed is for the WKB wave function to have the form

$$\psi^{WKB}(x) = \psi^{(0)}(x)/[\sqrt{p(x)} + \hbar f(x,\hbar)],$$

where $f(x,\hbar)$ is an analytic function of x and $\hbar$. It is easy to check that requiring $\psi^{WKB}(x)$ and its derivatives to be continuous amounts to the procedure of matching the value and slope of $\psi^{(0)}(x)$ at the classical turning point x_R, which justifies the simple approach.

8.2   Some General Comments on WKB Theory

There are two aspects to WKB theory. The first is its ability to accurately determine the energy eigenvalues and the second is its ability to describe the tunneling rate. These are not totally independent since the spectrum is also related to an analytic continuation of the scattering amplitude. Here we will concentrate on the validity of WKB theory for the spectrum. WKB theory should give good results if the turning points are several wave lengths apart or if n is large compared to unity. By now it has been tested for several potentials and one finds that for many of them even for low values of n it yields moderately accurate eigenvalues. For additional accuracy, it is necessary to consider second and higher order corrections in $\hbar$. In fact, this has been done. For example, it has been shown that to $O(\hbar^2)$, the WKB quantization condition (8.13) is modified to

$$\int_{x_L}^{x_R} dx \sqrt{2m[E - V(x)]} - \frac{\hbar^2}{24\sqrt{2m}} \frac{d}{dE} \int_{x_L}^{x_R} dx \frac{V''(x)}{\sqrt{E - V(x)}} = (n + 1/2)\hbar\pi \,.$$
(8.20)

In the special case of the one dimensional harmonic oscillator and the Morse potential, it turns out that the lowest order WKB approximation (8.13) is in fact exact (see the problems below) and further, the higher order corrections are all zero.

The WKB approximation can also be applied to three-dimensional problems with spherical symmetry by applying the WKB formalism to the reduced radial Schrödinger equation

$$\frac{\hbar^2}{2m} \frac{d^2 R}{dr^2} + [E - V_{eff}(r)]R(r) = 0 \,,$$
(8.21)

where the effective potential is

$$V_{eff}(r) = V(r) + \frac{l(l + 1)\hbar^2}{2mr^2} \,.$$
(8.22)

In view of the wrong behavior of the WKB reduced radial wave function at the origin, it was suggested by Langer that in the effective potential (8.22), $l(l + 1)$ be replaced by $(l + \frac{1}{2})^2$. This is popularly known in the literature as the Langer correction. It turns out that with this Langer correction, the lowest order WKB quantization condition reproduces the exact spectrum in the case of the Coulomb as well as the oscillator potentials (see the problems below). It may however be noted here that the Langer correction needs modification at each order of approximation.

8.3 Tunneling Probability in the WKB Approximation

Tunneling is one of the most striking consequence of quantum mechanics which has no classical analogue. There are numerous applications of this phenomenon starting from α decay. In most cases an exact computation of the tunneling probability is not possible and the WKB approximation has proved useful. For simplicity we again consider a symmetric potential in one dimension but now instead of a potential well we are considering a barrier. Let us assume that a particle of energy E is incident from the left and that E is less than the top of the potential barrier. Let x_L and x_R denote the two turning points. By exactly following the treatment given above for the potential well case, we can write down the WKB wave functions in the various regions. Note however that now in region I ($x_L \leq x \leq x_R$) one will have exponentially decaying and growing wave functions while in region II ($x < x_L$) and III ($x > x_R$) one will have oscillating wave functions. As before, these WKB wave functions will not be valid near the two turning points and one has to make proper use of the connection formulae to match the WKB solutions in the three regions. This has been discussed in great detail in the literature and it has been shown that the transmission and reflection probabilities are given by

$$| T |^2 = \frac{1}{1 + e^{2K}} \; ; \; | R |^2 = \frac{e^{2K}}{1 + e^{2K}} , \qquad (8.23)$$

where

$$K = \frac{1}{\hbar} \int_{x_L}^{x_R} dx \sqrt{2m[V(x) - E]} . \qquad (8.24)$$

Now let us discuss the accuracy of the WKB approximation (8.23) for the tunneling probability. In the classical limit ($\hbar \to 0$), $T \to 0$, i.e. there is indeed no barrier penetration. Further, as expected, $| R |^2 + | T |^2$ is indeed equal to 1. Besides, in the case of the inverted oscillator and the inverted Morse potential, the WKB approximation (8.23) for the tunneling probability is in fact exact. However, for other potentials, the WKB approximation is only moderately good. Another way to test the WKB approximation is to examine the poles of the transmission probability function $|T|^2$ analytically continued to the case of the inverted potential (well) and compare it with the exact bound state spectrum. It turns out that the poles of $|T|^2$ as given by eq. (8.23) indeed give the exact bound state spectrum in the case of the harmonic oscillator as well as the Morse potential while for all other cases it does not reproduce the exact spectrum.

8.4 SWKB Quantization Condition for Unbroken Supersymmetry

In the previous sections, we have reviewed the semiclassical WKB method. Combining the ideas of SUSY with the lowest order WKB method, Comtet, Bandrauk and Campbell obtained the lowest order SWKB quantization condition for unbroken SUSY and showed that it yields energy eigenvalues which are not only accurate for large quantum numbers n but which are also exact for the ground state ($n = 0$). We shall now discuss this in detail.

For the potential $V_1(x)$ corresponding to the superpotential $W(x)$, the lowest order WKB quantization condition (8.13) takes the form

$$\int_{x_L}^{x_R} \sqrt{2m\left[E_n^{(1)} - W^2(x) + \frac{\hbar}{\sqrt{2m}}W'(x)\right]} \, dx = (n + 1/2)\hbar\pi . \quad (8.25)$$

Let us assume that the superpotential $W(x)$ is formally $O(\hbar^0)$. Then, the W' term is clearly $O(\hbar)$. Therefore, expanding the left hand side in powers of $\hbar$ gives

$$\int_a^b \sqrt{2m[E_n^{(1)} - W^2(x)]} \, dx + \frac{\hbar}{2}\int_a^b \frac{W'(x) \, dx}{\sqrt{E_n^{(1)} - W^2(x)}} + \ldots = (n + 1/2)\hbar\pi ,$$

$$(8.26)$$

where a and b are the turning points defined by $E_n^{(1)} = W^2(a) = W^2(b)$. The $O(\hbar)$ term in eq. (8.26) can be integrated easily to yield

$$\frac{\hbar}{2}\sin^{-1}\left[\frac{W(x)}{\sqrt{E_n^{(1)}}}\right]_a^b . \quad (8.27)$$

In the case of unbroken SUSY, the superpotential $W(x)$ has opposite signs at the two turning points, that is

$$-W(a) = W(b) = \sqrt{E_n^{(1)}} . \qquad (8.28)$$

For this case, the $O(\hbar)$ term in (8.27) exactly gives $\hbar\pi/2$, so that to leading order in $\hbar$ the SWKB quantization condition when SUSY is unbroken is

$$\int_a^b \sqrt{2m[E_n^{(1)} - W^2(x)]} \, dx = n\hbar\pi, \quad n = 0, 1, 2, \dots . \qquad (8.29)$$

Proceeding in the same way, the SWKB quantization condition for the potential $V_2(x)$ turns out to be

$$\int_a^b \sqrt{2m[E_n^{(2)} - W^2(x)]} \, dx = (n+1)\hbar\pi, \quad n = 0, 1, 2, \dots . \qquad (8.30)$$

Some remarks are in order at this stage.

(i) For $n = 0$, the turning points a and b in eq. (8.29) are coincident and $E_0^{(1)} = 0$. Hence the SWKB condition is exact by construction for the ground state energy of the potential $V_1(x)$.

(ii) On comparing eqs. (8.29) and (8.30), it follows that the lowest order SWKB quantization condition preserves the SUSY level degeneracy i.e. the approximate energy eigenvalues computed from the SWKB quantization conditions for $V_1(x)$ and $V_2(x)$ satisfy the exact degeneracy relation $E_{n+1}^{(1)} = E_n^{(2)}$.

(iii) Since the lowest order SWKB approximation is not only exact, as expected for large n, but is also exact by construction for $n = 0$, hence, unlike the ordinary WKB approach, the SWKB eigenvalues are constrained to be accurate at both ends, at least when the spectrum is purely discrete. One can thus reasonably expect better results than the WKB scheme even when n is neither small nor very large.

(iv) For spherically symmetric potentials, unlike the conventional WKB approach, in the SWKB case one obtains the correct threshold behavior without making any Langer-like correction. This happens because, in this approach

$$S_0 \sim (E - W^2)^{1/2} \overset{r \to 0}{\sim} -i\hbar(l + 1)/r , \qquad (8.31)$$

so that

$$\psi(r) \sim \exp\left[\frac{i}{\hbar} \int^r S_0 dr\right] \overset{r \to 0}{\sim} r^{l+1} . \qquad (8.32)$$

One can show that even after including higher order correction terms like $S_1, S_2, ...$, the SWKB wave function continues to behave like r^{l+1} as $r \to 0$ to all orders in $\hbar$, i.e. the SWKB formalism contains the correct threshold behavior in a natural way.

8.5 Exactness of the SWKB Condition for Shape Invariant Potentials

In order to determine the accuracy of the SWKB quantization condition as given by eq. (8.29), researchers first obtained the SWKB bound state spectra of several analytically solvable potentials. Remarkably they found that the lowest order SWKB condition gives the exact eigenvalues for all SIPs with translation! Let us now prove this result.

Recall that the shape invariance condition eq. (4.1) on the partner potentials is

$$V_2(x, a_1) = V_1(x, a_2) + R(a_1),$$

where a_1 is a set of parameters, a_2 is a function of a_1 (say $a_2 = f(a_1)$) and the remainder $R(a_1)$ is independent of x.

In Chapter 4, we showed using factorization and the Hamiltonian hierarchy that the general expression for the s'th Hamiltonian was given by eq. (4.3)

$$H_s = -\frac{\hbar^2}{2m}\frac{d^2}{dx^2} + V_1(x, a_s) + \sum_{k=1}^{s-1} R(a_k),$$

where $a_s = f^{s-1}(a_1)$ i.e. the function f applied $s - 1$ times.

The proof of the exactness of the bound state spectrum eq. (4.6) in the lowest order SWKB approximation now follows from the fact that the SWKB condition (8.29) preserves (a) the level degeneracy and (b) a vanishing ground state energy eigenvalue. For the hierarchy of Hamiltonians $H^{(s)}$ as given by eq. (4.3), the SWKB quantization condition takes the form

$$\int \sqrt{2m\left[E_n^{(s)} - \sum_{k=1}^{s-1} R(a_k) - W^2(a_s; x)\right]}\, dx = n\hbar\pi. \qquad (8.33)$$

Now, since the SWKB quantization condition is exact for the ground state energy when SUSY is unbroken, hence

$$E_0^{(s)} = \sum_{k=1}^{s-1} R(a_k) \qquad (8.34)$$

must be exact for Hamiltonian $H^{(s)}$ as given by eq. (8.33). One can now go back in sequential manner from $H^{(s)}$ to $H^{(s-1)}$ to $H^{(2)}$ and $H^{(1)}$ and use the fact that the SWKB method preserves the level degeneracy $E_{n+1}^{(1)} = E_n^{(2)}$. On using this relation n times, we find that for all SIPs, the lowest order SWKB condition gives the exact energy eigenvalues.

This is a substantial improvement over the usual WKB formula eq. (8.13) which is not exact for most SIPs. Of course, one can artificially restore exactness by ad hoc Langer-like corrections. However, such modifications are unmotivated and have different forms for different potentials. Besides, even with such corrections, the higher order WKB contributions are non-zero for most of these potentials.

What about the higher order SWKB contributions? Since the lowest order SWKB energies are exact for shape invariant potentials, it would be nice to check that higher order corrections vanish order by order in $\hbar$. By starting from the higher order WKB formalism, one can readily develop the higher order SWKB formalism. It has been explicitly checked for all known SIPs (with translation) that up to $O(\hbar^6)$ there are indeed no corrections.. This result can be extended to all orders in $\hbar$.

We proved above that the lowest order SWKB approximation reproduces the exact bound state spectrum of any SIP. This statement has indeed been explicitly checked for all known SIPs with translation i.e. solutions of the shape invariance condition involving a translation of parameters $a_2 = a_1+$ constant. However, a few years ago it has was shown that the above statement is not true for the newly discovered class of SIPs discussed in Chapter 4, for which the parameters a_2 and a_1 are related by scaling $a_2 = qa_1$. This is because for those potentials, W is not explicitly known except as a power series which mixes powers of $\hbar$ so that the superpotential is intrinsically $\hbar$-dependent and hence the derivation given above is no longer valid for these SIPs.

8.6   Comparison of the SWKB and WKB Approaches

Let us now compare the merits of the WKB and SWKB methods. For potentials for which the ground state wave function (and hence the superpotential W) is not known, clearly the WKB approach is preferable, since one cannot directly make use of the SWKB quantization condition (8.29). On the other hand, we have already seen that for shape invariant potentials, SWKB is clearly superior. An obvious interesting question is to compare WKB and SWKB for potentials which are not shape invariant but for whom the ground state wave function is known. One choice which readily springs to mind is the Ginocchio potential given by

$$V(x) = (1 - y^2) \left\{ -\lambda^2 \nu(\nu + 1) + \frac{(1 - \lambda^2)}{4} [2 - (7 - \lambda^2)y^2 + 5(1 - \lambda^2)y^4] \right\}$$

$$(8.35)$$

where y is related to the independent variable x by

$$\frac{dy}{dx} = (1 - y^2)[1 - (1 - \lambda^2)y^2] . \qquad (8.36)$$

Here the parameters ν and λ measure the depth and shape of the potential respectively. The corresponding superpotential is

$$W(x) = (1 - \lambda^2)y(y^2 - 1)/2 + \mu_0 \lambda^2 y , \qquad (8.37)$$

where μ_n is given by Ginocchio

$$\mu_n \lambda^2 = \sqrt{[\lambda^2(\nu + 1/2)^2 + (1 - \lambda^2)(n + 1/2)^2]} - (n + 1/2) , \qquad (8.38)$$

and the bound state energies are

$$E_n = -\mu_n^2 \lambda^4, \quad n = 0, 1, 2, \ldots . \qquad (8.39)$$

For the special case $\lambda = 1$, one has the Rosen-Morse potential, which is shape invariant. The spectra of the Ginocchio potential using both the WKB and SWKB quantization conditions have been computed. The results are shown in Table 8.1.

In general, neither semiclassical method gives the exact energy spectrum. The only exception is the shape invariant limit $\lambda = 1$, in which case the SWKB results are exact, as expected. Also, for $n = 0, 1$ the SWKB values are consistently better, but there is no clear cut indication that SWKB

Table 8.1 Comparison of the lowest order WKB and SWKB predictions for the bound state spectrum of the Ginocchio potential for different values of the parameters λ, ν and several values of the quantum number n. The exact answer is also given. Units corresponding to $\hbar = 2m = 1$ are used throughout.

n	$\lambda = 0.5,$ WKB	$\nu = 5.5$ SWKB	Exact	$\lambda = 6.25,$ WKB	$\nu = 5.5$ SWKB	Exact
0	−6.19	−6.41	−6.41	−1372.28	−1359.61	−1359.61
1	−3.08	−3.17	−3.13	−1228.16	−1212.70	−1213.84
2	−1.43	−1.47	−1.44	−1012.52	−999.63	−1003.70
3	−0.58	−0.60	−0.58	−733.59	−727.72	−737.62
5	−0.02	−0.02	−0.02	−55.27	−70.59	−109.50

n	$\lambda = 0.5,$ WKB	$\nu = 10.5$ SWKB	Exact	$\lambda = 6.25,$ WKB	$\nu = 10.5$ SWKB	Exact
0	−24.87	−25.17	−25.17	−4659.88	−4648.62	−4648.61
1	−17.09	−17.27	−17.23	−4452.74	−4438.32	−4438.80
2	−11.57	−11.68	−11.63	−4174.48	−4158.48	−4159.43
3	−7.70	−7.77	−7.73	−328.49	−3812.57	−3815.65
5	−3.15	−3.18	−3.15	−2949.60	−2978.47	−2947.70

results are always better. This example, as well as several other potentials (including the one parameter family of potentials which are strictly isospectral to the SIPs with translation) studied in the literature indicate that by and large, SWKB does better than WKB in case the ground state wave function and hence the superpotential W is known. These studies also support the conjecture that shape invariance is perhaps a necessary condition so that the lowest order SWKB reproduce the exact bound state spectrum.

8.7 SWKB Quantization Condition for Broken Supersymmetry

The derivation of the lowest order SWKB quantization condition (8.29) for the case of unbroken SUSY is given above [(8.25) to (8.29)]. For the case of broken SUSY, the same derivation applies until one examines the $O(\hbar)$ term in eq. (8.27). Here, for broken SUSY, one has

$$W(a) = W(b) = \sqrt{E_n^{(1)}} \, , \tag{8.40}$$

and the $O(\hbar)$ term in (8.27) exactly vanishes. So, to the leading order in $\hbar$, the SWKB quantization condition for broken SUSY (BSWKB) is

$$\int_a^b \sqrt{2m[E_n^{(1)} - W^2(x)]}\, dx = (n + 1/2)\hbar\pi, \quad n = 0, 1, 2, \dots . \quad (8.41)$$

As before, it is easy to obtain the quantization condition which includes higher orders in $\hbar$ and to test how well the broken SWKB condition works for various specific examples. As in the unbroken SUSY case, it is found that the lowest order BSWKB quantization condition also reproduces the exact spectra for SIPs with translation of parameters. For potentials which are not analytically solvable, the results using eq. (8.41) are usually better than the standard WKB computations.

8.8   Tunneling Probability in the SWKB Approximation

In this section we shall show that in certain respects, the SWKB approximation does even better than the WKB approximation as far as the computation of the tunneling probability $\mid T \mid^2$ is concerned. In particular, recall that whereas the WKB expression for $\mid T \mid^2$ is exact in the case of the inverted oscillator and the inverted Morse potential, in the other cases not only is the expression for $\mid T \mid^2$ not exact but even the poles of $\mid T \mid^2$ analytically continued to the inverted potential (well instead of barrier) do not reproduce the correct bound state energy eigenvalues except for the harmonic oscillator and the Morse potentials. For the SWKB method, it is again true that except for the two special cases mentioned above, one does not get the exact expression for $\mid T \mid^2$. Nevertheless the poles of $\mid T \mid^2$, analytically continued to the inverted potential (which is now a well), do reproduce the exact bound state spectrum for SIPs with translation.

Consider a symmetric potential barrier in one dimension. We start from the WKB expression for $\mid T \mid^2$ as given by

$$\mid T \mid^2 = \frac{1}{1 + e^{2K}}, \quad (8.42)$$

where

$$K = \frac{1}{\hbar} \int_{x_L}^{x_R} dx \sqrt{2m[V(x) - E]}. \quad (8.43)$$

Let us apply this formula to the partner potentials

$$V_{1,2}(x) = W^2(x) \mp \frac{\hbar}{\sqrt{2m}} W'(x),$$ (8.44)

where $W(x)$ is the analytically continued superpotential for the barriers. Clearly, for the partner potentials, K takes the form

$$K^{(1,2)} = \frac{\sqrt{2m}}{\hbar} \int_a^b dx \sqrt{W^2(x) - E}$$

$$\mp \frac{1}{2} \int_{b_L}^{b_R} \frac{dW}{\sqrt{W^2(x) - E}},$$ (8.45)

where b, a are the turning points obtained from

$$W^2(b) = E = W^2(a).$$ (8.46)

The value of the second integral of eq. (8.45) turns out to be $-i\pi$ and hence the expression for the transmission probability for the SUSY partner potentials is given by

$$|T_{SWKB}^{(1,2)}|^2 = \frac{1}{1 + e^{2K^{(1,2)}}},$$ (8.47)

where

$$K^{(1,2)} = \frac{\sqrt{2m}}{\hbar} \int_a^b p^{(1,2)}(x) dx \pm \frac{i\pi}{2},$$ (8.48)

and

$$p^{(1,2)}(x) = \sqrt{W^2(x) - E^{(1,2)}}.$$ (8.49)

Now, under the change $p^{(1,2)} \to ip^{(1,2)}$, $T^{(1,2)}$ have poles in the complex energy plane and they precisely give the SWKB quantization conditions for $V^{(1,2)}$ as given by eqs. (8.29) and (8.30) which have been shown to yield the exact bound state spectrum for all SIPs with translation. In this way, we see that the SWKB expression for $|T|^2$ when analytically continued, yields the exact bound state spectrum for all SIP with translation. It must however be emphasized that as far as the expression for $|T|^2$ itself is concerned, SWKB is in general inexact, the two exceptions being the inverted oscillator and the inverted Morse potential. However, as in the bound state spectrum case, one finds that on the whole, the SWKB expression for $|T|^2$ gives better results compared to the corresponding WKB answer.

References

(1) C.M. Bender and S.A. Orszag, *Advanced Mathematical Methods for Scientists and Engineers*, McGraw-Hill (1978).

(2) N. Fröman and P.O. Fröman, *JWKB Approximation*, North Holland (1960).

(3) J.L. Dunham, *The Wentzel-Brillouin-Kramers Method of Solving the Wave Function*, Phys. Rev. **41** (1932) 713-720.

(4) J.B. Krieger and C. Rosenzweig, *Application of a Higher Order WKB Approximation to Radial Problems*, Phys. Rev. **164** (1967) 171-173; C. Rosenzweig and J.B. Krieger, *Exact Quantization Conditions* J. Math. Phys. **9** (1968) 849-860; J.B. Krieger, M. Lewis and C. Rosenzweig, *Use of the WKB Method for Obtaining Energy Eigenvalues*, J. Chem. Phys. **47** (1967) 2942-2945.

(5) J.J. Sakurai, *Modern Quantum Mechanics*, Addison-Wesley (1994); R. Shankar, *Principles of Quantum Mechanics*, Plenum Press (1980) E. Merzbacher, *Quantum Mechanics*, 2nd ed., Wiley (1970); L. Schiff, *Quantum Mechanics*, 3rd ed., McGraw-Hill (1968).

(6) R.E. Langer, *On the Connection Formulas and the Solutions of the Wave Equation*, Phys. Rev. **51** (1937) 669-676; S. C. Miller and R. H. Good, *A WKB-Type Approximation to the Schrödinger Equation*, Phys. Rev. **91** (1953) 174-179.

(7) U.P. Sukhatme and M.N. Sergeenko, *Semiclassical Approximation for Periodic Potentials*, E-Print archive quant-ph/9911026 (1999).

(8) A. Comtet, A. Bandrauk and D.K. Campbell, *Exactness of Semiclassical Bound State Energies for Supersymmetric Quantum Mechanics*, Phys. Lett. **B150** (1985) 159-162.

(9) R. Dutt, A. Khare and U. Sukhatme, *Exactness of Supersymmetric WKB Spectrum for Shape Invariant Potentials*, Phys. Lett. **B181** (1986) 295-298.

(10) R. Adhikari, R. Dutt, A. Khare and U.P. Sukhatme, *Higher Order WKB Approximations in Supersymmetric Quantum Mechanics*, Phys. Rev. **A38** (1988) 1679-1686.

(11) R. Dutt, A. Khare and U. Sukhatme, *Supersymmetry-Inspired WKB Approximation in Quantum Mechanics*, Am. Jour. Phys. **59** (1991) 723-727.

(12) A.Khare and Y.P. Varshni, *Is Shape invariance Also Necessary for Lowest Order Supersymmetricd WKB to be Exact?*, Phys. Lett

A142 (1989) 1-4.

(13) A. Inomata and G. Junker, *Lectures on Path Integration*, Editors: H. Cerdeira et al., World Scientific (1992).

(14) R. Dutt, A. Gangopadhyaya, A. Khare, A. Pagnamenta and U.P. Sukhatme, *Semiclassical Approach to Quantum Mechanical Problems with Broken Supersymmetry*, Phys. Rev. **A48** (1993) 1845-1853.

Problems

1. Work out the energy spectra of the harmonic oscillator and the Morse potentials using the WKB approximation and show that they are exact. In both cases work out the $O(\hbar^2)$ correction using the formula given in the text, and show that the corrections vanish. Show that the SWKB approximation for these potentials also gives exact energy eigenvalues.

2. For the potential $V(x) = A^2 - A(A+1)\operatorname{sech}^2 x$, show that the WKB approximation does not reproduce the exact spectrum. Compute the accuracy of the four eigenenergies corresponding to the case $A = 4$. Now repeat the calculation using the SWKB approximation, and show that it gives exact eigenenergies.

3. Compute the spectrum of the Coulomb potential $V(r) = -e^2/r$ using the WKB approximation with the appropriate $l(l+1)/2mr^2$ angular momentum term. Show that the inclusion of the Langer correction $l(l+1) \to (l+\frac{1}{2})^2$ leads to the exact spectrum. Finally, show that the SWKB approximation also gives the exact eigenenergies, thus removing the need for any ad hoc Langer-type corrections.

4. Consider the one-dimensional potential $V(x) = \alpha|x|^q$. Show that the WKB energy levels are given by

$$E_n = \left\{ \frac{(n + \frac{1}{2})h\alpha^{1/q}\Gamma(\frac{3}{2} + q^{-1})}{4(2m)^{1/2}\Gamma(\frac{3}{2})\Gamma(1 + q^{-1})} \right\}^{2q/(q+2)},$$

where $\Gamma(x)$ is the usual gamma function. Note that no power law potential exists which can make energy levels further apart asymptotically than the quadratic spacing of the infinite square well.

Chapter 9

Perturbative Methods for Calculating Energy Spectra and Wave Functions

The framework of supersymmetric quantum mechanics has been very useful in generating several new perturbative methods for calculating the energy spectra and wave functions for one dimensional potentials. Four such methods are described in this chapter.

In Secs. 9.1 and 9.2, we discuss two approximation methods (the variational method and the $\delta-$ expansion) for determining the wave functions and energy eigenvalues of the anharmonic oscillator making use of SUSY QM. Sec. 9.3 contains a description of a SUSY QM calculation of the energy splitting and rate of tunneling in a double well potential. The result is a rapidly converging series which is substantially better than the usual WKB tunneling formula. Finally, in Sec. 9.4, we describe how the large N expansion (N = number of spatial dimensions) used in quantum mechanics can be further improved by incorporating SUSY.

9.1 Variational Approach

The anharmonic oscillator potential $V(x) = gx^4$ is not an exactly solvable problem in quantum mechanics. To determine the superpotential one has to first subtract the ground state energy E_0 and solve the Riccati equation for $W(x)$:

$$V_1(x) = gx^4 - E_0 \equiv W^2 - W' . \qquad (9.1)$$

Once the ground state energy and the superpotential is known to some order of accuracy, one can then determine the partner potential and its ground state wave function approximately. Then, using the SUSY operator

$$A^+ \equiv -\frac{\hbar}{\sqrt{2m}}\frac{d}{dx} + W(x)$$

one can construct the first excited state of the anharmonic oscillator in the usual manner. Using the hierarchy of Hamiltonians discussed in Chapter 3, one can construct from the approximate ground state wave functions of the hierarchy and the approximate superpotentials W_n all the excited states of the anharmonic oscillator approximately.

First let us see how this works using a simple variational approach. For the original potential, we can determine the optimal Gaussian wave function quite easily. Assuming a (normalized) trial wave function of the form

$$\psi_{0v} = (\frac{2\beta}{\pi})^{1/4}e^{-\beta x^2} , \tag{9.2}$$

we obtain

$$< H >=< \frac{p^2}{2} + gx^4 >= \frac{\beta}{2} + \frac{3g}{16\beta^2} . \tag{9.3}$$

(In this section, we are taking $m = 1, \hbar = 1$ in order to make contact with published numerical results). Minimizing the expectation value of the Hamiltonian with respect to the parameter β yields

$$E_0 = (\frac{3}{4})^{4/3}g^{1/3} , \qquad \beta = (\frac{3}{4})^{1/3}g^{1/3} .$$

This is rather good for this crude approximation since the exact ground state energy of the anharmonic oscillator determined numerically is $E_0 = 0.66799g^{1/3}$ whereas $(\frac{3}{4})^{4/3} = 0.68142$. The approximate superpotential W resulting from this variational calculation is

$$W(x) = -\frac{\psi'_{0v}(x)}{\psi_{0v}(x)} = 2\beta x , \tag{9.4}$$

which leads, within the Gaussian approximation, to the potential

$$V_{1G} = 4\beta x^2 - 2\beta . \tag{9.5}$$

The (approximate) supersymmetric partner potential is now

$$V_{2G} = 4\beta x^2 + 2\beta . \tag{9.6}$$

Table 9.1 Comparison of the three lowest energy eigenvalues obtained by a variational method with the exact results.

Level	n	ρ	$\Delta(E)_{var}$	$\Delta(E)_{exact}$
0	1.183458	0.666721	0.669330	0.667986
1	0.995834	0.429829	1.727582	1.725658
2	1.000596	0.435604	2.316410	2.303151

Since V_{2G} differs from V_{1G} by a constant, the approximate ground state wave function for V_2 is also given by eq. (9.2). The approximate ground state energy of the second potential is now

$$< H_2 >=< \psi_{0v}|\frac{p^2}{2} + V_{2G}|\psi_{0v} >=< H_1 > +4\beta . \qquad (9.7)$$

Thus we find in the harmonic approximation that the energy difference between the ground state and the first excited state of the anharmonic oscillator is

$$E_1 - E_0 = 4\beta = 4(\frac{3}{4})^{1/3}g^{1/3} = 3.632g^{1/3} ,$$

which is to be compared with the exact numerical value of $1.726g^{1/3}$ as seen from the Table 9.1. This shows that the harmonic approximation breaks down rapidly when we consider the higher energy eigenstates of an anharmonic oscillator.

The approximate (unnormalized) first excited state wave function in this simple approximation is

$$\psi_1^{(1)} = [-\frac{1}{\sqrt{2}}\frac{d}{dx} + 2\beta x]\psi_{0v} \propto 4\beta xe^{-\beta x^2} . \qquad (9.8)$$

To obtain better accuracy it is necessary to extend the number of variational parameters. For the ground state wave function one does not want any nodes. A parameterization which allows one to perform all the integrals analytically in the determination of the trial Hamiltonian is a generalization of the Gaussian to the form:

$$\psi_0^{(k)} = N_k \exp\left[-\frac{1}{2}\left(\frac{x^2}{\rho_k}\right)^{n_k}\right] , \qquad N_k = \left[2\sqrt{\rho_k}\Gamma(1 + \frac{1}{2n_k})\right]^{-1/2} . \qquad (9.9)$$

Using this generalized form, we obtain much better agreement for the low lying eigenvalues and eigenfunctions as we shall demonstrate below. It is convenient in this case to first scale the Hamiltonian for the anharmonic oscillator,

$$H = -\frac{1}{2}\frac{d^2}{dx^2} + gx^4 , \tag{9.10}$$

by letting $x \to x/g^{1/6}$ and $H \to g^{1/3}H$. Then we find the ground state energy of the anharmonic oscillator and the variational parameters ρ_1 and n_1 by forming the functional

$$E_0(\rho_1,n_1) = \langle \psi_0^{(1)} | -\frac{1}{2}\frac{d^2}{dx^2} + x^4 | \psi_0^{(1)} \rangle . \tag{9.11}$$

Thus we first determine ρ_1 and n_1 by requiring

$$\frac{\partial E_0}{\partial \rho_1} = 0 , \qquad \frac{\partial E_0}{\partial n_1} = 0 . \tag{9.12}$$

Using the trial wavefunction (9.9), the energy functional for the anharmonic oscillator is given by

$$E_0(\rho_1,n_1) = \frac{n_1^2}{2\rho_1}\frac{\Gamma(2-\frac{1}{2n_1})}{\Gamma(\frac{1}{2n_1})} + \rho_1^2\frac{\Gamma(\frac{5}{2n_1})}{\Gamma(\frac{1}{2n_1})} . \tag{9.13}$$

Minimizing this expression, we obtain the following variational result:

$$E_0 = 0.66933, \quad n_1 = 1.18346, \quad \rho_1 = 0.666721 . \tag{9.14}$$

This ground state energy is to be compared with the exact numerical value of 0.667986 .

Let us now try to estimate the energy differences $E_n - E_{n-1}$ of the anharmonic oscillator. To that end, we consider the variational Hamiltonian

$$\bar{H}_{vk+1} = \frac{1}{2}A_{kv}A_{kv}^\dagger \tag{9.15}$$

which approximately determines these energy differences. Since the trial wave function for all ground states is given by eq. (9.9), the variational superpotential for all k is

$$W_{kv} = n_k|x|^{2n_k-1}(\rho_k)^{-n_k} . \tag{9.16}$$

We obtain the approximate energy splittings by minimizing the energy functional

$$\delta E_k(\rho_k,n_k) = \frac{1}{2}\langle \psi_0^{(k+1)} | -\frac{d^2}{dx^2} + W_{vk}^2 + W_{vk}' | \psi_0^{(k+1)} \rangle . \tag{9.17}$$

Performing the integrals one obtains the simple recursion relation:

$$\delta E_k(\rho_k, n_k) = \frac{n_k^2}{2\rho_k} \frac{\Gamma\left(2 - \frac{1}{2n_k}\right)}{\Gamma\left(\frac{1}{2n_k}\right)} + \frac{n_{k-1}^2}{2\rho_k} \left(\frac{\rho_k}{\rho_{k-1}}\right)^{2n_{k-1}} \frac{\Gamma\left(\frac{4n_{k-1}-1}{2n_k}\right)}{\Gamma\left(\frac{1}{2n_k}\right)}$$

$$+ \frac{n_{k-1}}{2\rho_k}(2n_{k-1} - 1) \left(\frac{\rho_k}{\rho_{k-1}}\right)^{n_{k-1}} \frac{\Gamma\left(\frac{2n_{k-1}-1}{2n_k}\right)}{\Gamma\left(\frac{1}{2n_k}\right)} . \tag{9.18}$$

One can perform the minimization in ρ analytically leaving one minimization to perform numerically.

The results for the variational parameters and for the energy differences are presented in Table 9.1 for the first three energy eigenvalues and compared with a numerical calculation, based on a shooting method. For these low lying states, this variational method is more accurate than first order WKB results. However for $n \geq 4$ first order WKB becomes more accurate than this variational calculation. A more recent calculation using more variational parameters (polynomials times exponentials) gave energy eigenvalues for these low lying states accurate to .1% .

9.2 SUSY δ Expansion Method

In this section we consider the anharmonic oscillator as an analytic continuation from the harmonic oscillator in the parameter controlling the anharmonicity. We introduce a novel expansion method based on introducing a perturbation parameter δ which describes the degree' of nonlinearity of an anharmonic oscillator. We will follow the ideas of Appendix C in doing our perturbation theory by assuming that both the ground state energy as well as the superpotential $W(x)$ have an expansion in the perturbation parameter (here δ as opposed to the coupling constant version in Appendix C). We will make a slight change in notation here to conform with the published literature. We will choose the Hamiltonian for the quartic anharmonic oscillator as before, but introduce a new variable $V_1(x)$ to be twice the usual potential. Thus, we have for the anharmonic oscillator:

$$H = \frac{p^2}{2} + gx^4 \equiv \frac{p^2}{2} + \frac{V_1(x)}{2} . \tag{9.19}$$

On introducing a mass scale parameter M, and an anharmonicity parameter δ, $V_1(x)$ has the form

$$V_1(x) = M^{2+\delta}x^{2+2\delta} - C(\delta) \equiv W^2(x,\delta) - W' , \qquad (9.20)$$

where C is the ground state energy of the anharmonic oscillator. C is subtracted as usual from the potential so that it can be factorized. The harmonic oscillator had $\delta = 0$ and $M = m$. As we increase δ to one we reach the quartic anharmonic oscillator with the identification $M = (2g)^{1/3}$. To approximately determine $W(x)$ from $V_1(x)$ we assume that both $W(x)$ and $V_1(x)$ have a Taylor series expansion in δ. Thus we write:

$$V_1(x) = M^2x^2 \sum_{n=0}^{\infty} \frac{\delta^n[\ln(Mx^2)]^n}{n!} - \sum_{n=0}^{\infty} 2E_n\delta^n , \qquad (9.21)$$

where E_n corresponds to the coefficient of δ^n in the Taylor series expansion of the ground state energy.

We assume

$$W(x) = \sum_{n=0}^{\infty} \delta^n W_{(n)}(x) , \qquad (9.22)$$

and insert these expressions in eq. (9.20) and match terms order by order. At lowest order in δ the problem reduces to the supersymmetric harmonic oscillator. We have:

$$W_0^2 - W_0' = M^2x^2 - 2E_0 , \qquad (9.23)$$

whose solution is

$$W_0(x) = Mx , \qquad E_0 = \frac{1}{2}M . \qquad (9.24)$$

To the next order we have the differential equation:

$$\frac{dW_1}{dx} - 2W_1W_0 = -M^2x^2\ln(Mx^2) + 2E_1 , \qquad (9.25)$$

which is to be solved with the boundary condition $W_n(0) = 0$. The order δ contribution to the energy eigenvalue E_1 is determined by requiring that the ground state wave function be square integrable. Solving for W_1 we obtain

$$W_1(x) = -e^{Mx^2}\int_0^x dy\, e^{-My^2}[M^2y^2\ln(My^2) - 2E_1] . \qquad (9.26)$$

To first order in δ the ground state wave function is now:

$$\psi_0(x) = e^{-Mx^2}[1 - \delta \int_0^x dy W_1(y)] .$$

Imposing the condition that ψ_0 vanishes at infinity, we obtain:

$$E_1 = \frac{1}{4}M\psi(3/2) , \quad \psi(x) = \Gamma'(x)/\Gamma(x) . \tag{9.27}$$

Writing $M = (2g)^{1/3}$, we find that the first two terms in the δ expansion for the ground state energy are

$$E = \frac{1}{2}(2g)^{1/3}[1 + \frac{1}{2}\psi(3/2)\delta] . \tag{9.28}$$

At $\delta = 1$, we get for the ground state energy of the quartic anharmonic oscillator

$$E = 0.6415g^{1/3} . \tag{9.29}$$

as opposed to the exact numerical value $E = 0.667986g^{1/3}$. A more accurate determination of the ground state energy can be obtained by calculating up to order δ^2 and then analytically continuing in δ using Padé approximants. Using standard SUSY methods we can also calculate all the excited states of the anharmonic oscillator in a δ expansion about the Harmonic oscillator result. The method can also be extended to perturbing about any shape invariant potential. For determining the energy levels of the anharmonic oscillator the variational method is simpler and more accurate than the δ expansion method.

9.3 Supersymmetry and Double Well Potentials

Supersymmetric quantum mechanics has been profitably used to obtain a novel perturbation expansion for the probability of tunneling in a double well potential. Since double wells are widely used in many areas of physics and chemistry, this expansion has found many applications ranging from condensed matter physics to the computation of chemical reaction rates. In what follows, we shall restrict our attention to symmetric double wells, although an extension to asymmetric double wells is relatively straightforward.

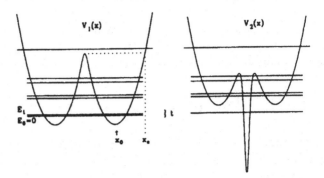

Fig. 9.1   A "deep" symmetric double well potential $V_1(x)$ with minima at $x = \pm x_0$ and
its supersymmetric partner potential $V_2(x)$.

Usually, in most applications the quantity of interest is the energy difference $t \equiv E_1 - E_0$ between the lowest two eigenstates, and corresponds to
the tunneling rate through the double-well barrier. The quantity t is often
small and difficult to calculate numerically, especially when the potential
barrier between the two wells is large. Here, we show how SUSY facilitates
the evaluation of t. Indeed, using the supersymmetric partner potential
$V_2(x)$, we obtain a systematic, highly convergent perturbation expansion
for the energy difference t. The leading term is more accurate than the
standard WKB tunneling formula, and the magnitude of the nonleading
terms gives a reliable handle on the accuracy of the result.

First, we briefly review the standard approach for determining t in the
case of a symmetric, one-dimensional double well potential, $V_1(x)$, whose
minima are located $x = \pm x_0$. We define the depth, D, of $V_1(x)$ by $D \equiv
V_1(0) - V_1(x_0)$. An example of such a potential is shown in Fig. 9.1

For sufficiently deep wells, the double-well structure produces closely
spaced pairs of energy levels lying below $V_1(0)$. The number of such pairs
n, can be crudely estimated from the standard WKB bound-state formula
applied to $V_1(x)$ for $x > 0$:

$$n\pi = \int_0^{x_c} [V_1(0) - V_1(x)]^{1/2} dx \ , \tag{9.30}$$

where x_c is the classical turning point corresponding to energy $V_1(0)$ and we
have chosen units where $\hbar = 2m = 1$. We shall call a double-well potential

"shallow" if it can hold at most one pair of bound states, i.e., $n \leq 1$. In contrast, a "deep" potential refers to $n \geq 2$.

The energy splitting t of the lowest-lying pair of states can be obtained by a standard argument. Let $\chi(x)$ be the normalized eigenfunction for a particle moving in a single well whose structure is the same as the right-hand well of $V_1(x)$ (i.e., $x > 0$). If the probability of barrier penetration is small, the lowest two eigenfunctions of the double-well potential $V_1(x)$ are well approximated by

$$\psi_{0,1}^{(1)}(x) = [\chi(x) \pm \chi(-x)]/\sqrt{2} . \tag{9.31}$$

By integration of Schrödinger's equation for the above eigenfunctions, it can be shown that

$$t \equiv E_1 - E_0 = 4\chi(0)\chi'(0) , \tag{9.32}$$

where the prime denotes differentiation with respect to x. This result is accurate for "deep" potentials, but becomes progressively worse as the depth decreases. Use of WKB wave functions in eq. (9.32) yields the standard result:

$$t_{\text{WKB}} = \{[2V_1''(x_0)]^{1/2}/\pi\} \exp\left(-2 \int_0^{x_0} [V_1(x) - V_1(x_0)]^{1/2} dx\right) . \tag{9.33}$$

Using the supersymmetric formulation of quantum mechanics for a given Hamiltonian, $H_1 = -d^2/dx^2 + V_1(x)$, and its zero-energy ground state wave function $\psi_0^{(1)}$, we know that the supersymmetric partner potential $V_2(x)$ is given by

$$\begin{aligned} V_2(x) &= V_1(x) - 2(d/dx)(\psi_0'/\psi_0) \\ &= -V_1(x) + 2(\psi_0'/\psi_0)^2 . \end{aligned} \tag{9.34}$$

(Here and in what follows we are using ψ_0 for $\psi_0^{(1)}$.)

Alternatively, in terms of the superpotential $W(x)$ given by $W(x) = -\psi_0'/\psi_0$ we can write

$$V_{2,1}(x) = W^2(x) \pm dW/dx . \tag{9.35}$$

From the discussion of unbroken SUSY in previous chapters, we know that the energy spectra of the potentials V_2 and V_1 are identical, except for the ground state of V_1 which is missing from the spectrum of V_2. Hence, for the double-well problem, we see that if $V_1(x)$ is "shallow" (i.e., only the lowest two states are paired), then the spectrum of V_2 is well separated. In this case, V_2 is relatively structureless and simpler than V_1. Thus, not surprisingly, for the shallow potentials, the use of SUSY simplifies the evaluation of the energy difference t. In contrast, let us now consider the case of a deep double well as shown in Fig. 9.1. Here, the spectrum of V_2 has a single unpaired ground state followed by paired excited states. In order to produce this spectrum, V_2 has a double-well structure together with a sharp "$\delta-$ function like" dip at $x = 0$. This central dip produces the unpaired ground state, and becomes sharper as the potential $V_1(x)$ becomes deeper.

As a concrete example, we consider the class of potentials whose ground state wave function is the sum of two Gaussians, centered around $\pm x_0$,

$$\psi_0(x) \sim e^{-(x-x_0)^2} + e^{-(x+x_0)^2} . \tag{9.36}$$

The variables x and x_0 have been chosen to be dimensionless. The corresponding superpotential $W(x)$, and the two supersymmetric partner potentials $V_1(x)$ and $V_2(x)$, are given respectively by

$$W(x) = 2[x - x_0 \tanh(2xx_0)] , \tag{9.37}$$

$$V_{1,2}(x) = 4[x - x_0 \tanh(2xx_0)]^2 \mp 2[1 - 2x_0^2 \mathrm{sech}^2(2xx_0)] . \tag{9.38}$$

The minima of $V_1(x)$ are located near $\pm x_0$ and the well depth (in the limit of large x_0) is $D \simeq 4x_0^2$. We illustrate the potentials $V_1(x)$ and $V_2(x)$ in Fig. 9.2 for the two choices $x_0 = 1.0$ and $x_0 = 2.5$. We see that in the limit of large x_0, for both $V_1(x)$ and $V_2(x)$, the wells become widely separated and deep and that $V_2(x)$ develops a strong central dip.

The asymptotic behavior of the energy splitting, t, in the limit $x_0 \to \infty$ can be calculated from eq. (9.32), with $\chi(x)$ given by one of the (normalized) Gaussians in eq. (9.36). We find that

$$t \to 8x_0(2/\pi)^{1/2} e^{-2x_0^2} . \tag{9.39}$$

The same result can be obtained by observing that $V_1(x) \to 4(|x| - x_0)^2$ as $x_0 \to \infty$. This potential has a well known analytic solution, which involves solving the parabolic cylindrical differential equation. After carefully handling the boundary conditions, one obtains the separation of the lowest two energy levels to be $8x_0(2/\pi)^{1/2} \exp(-2x_0^2)$, in agreement with eq. (9.39).

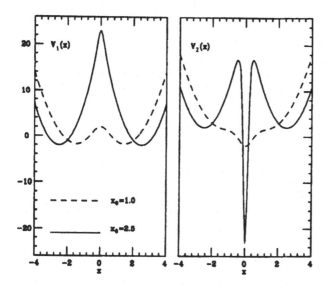

Fig. 9.2 Supersymmetric partner potentials $V_1(x)$ and $V_2(x)$ corresponding to two choices of the parameter x_0 for the potentials given in eq. (9.38).

We now turn to the evaluation of t by determining the ground state energy of the supersymmetric partner potential $V_2(x)$. In general, since $V_2(x)$ is not analytically solvable, we must solve an approximate problem and calculate the corrections perturbatively. We will first show that a close approximation to the potential can be found by studying the non-normalizable solution to Schrödinger equation for the potential $V_2(x)$. Using the fact that

$$W(x) = -\frac{\psi_0'}{\psi_0}$$

and that

$$V_2(x) = W^2 + W'$$

, it is straightforward to show that the wave function

$$\bar{\psi}(x) = \frac{1}{\psi_0(x)} \tag{9.40}$$

is a zero energy solution of the Schrödinger equation for the potential $V_2(x)$.

Since t is small, we expect this solution to be an excellent approximation to the correct eigenfunction for small values of x. However, $1/\psi_0$ is not normalizable and hence is not acceptable as a starting point for perturbation theory. One possibility is to regularize the behavior artificially at large $|x|$. This procedure is cumbersome and results in perturbation corrections to the leading term which are substantial.

It turns out, that if we consider the second linearly independent solution of the Schrödinger equation related to $\bar{\psi}(x)$

$$\phi(x) = \frac{1}{\psi_0} \int_x^\infty \psi_0^2(x')dx', \quad x > 0, \tag{9.41}$$

and $\phi(x) = \phi(-x)$ for $x < 0$, then this wave function actually corresponds to a well defined zero-energy solution of the Schrödinger equation for a slightly different potential $V_0(x)$, namely

$$V_0(x) = V_2(x) - 4\psi_0^2(0)\delta(x). \tag{9.42}$$

We can now do standard perturbation perturbation theory about $V_0(x)$ to find the approximate solution for $V_2(x)$. In eq. (9.42) we have assumed that $\psi_0(x)$ is normalized. The wave function $\phi(x)$ is well behaved at $x = \pm\infty$ and closely approximates $1/\psi_0$ at small x. It already is an excellent approximation of the exact ground state wave function of $V_2(x)$ for all values of x. The derivative of $\phi(x)$ is continuous except at the origin, where, unlike the exact solution to $V_2(x)$, it has a discontinuity

$$\phi\mid_{x=+\epsilon} -\phi\mid_{x=-\epsilon} = -2\psi_0(0).$$

We can calculate the perturbative corrections to the ground state energy using $\Delta V = +4\psi_0^2(0)\delta(x)$ as the perturbation. Note that the coefficient multiplying the δ-function is quite small as a result of $\psi(0)$ being small so that we expect our perturbation series to converge rapidly. It may be noted that since $\psi(0)$ is the ground state wave function for a double well potential, its support is mostly near the minima of the potential and it is smallest at the origin and at $\pm\infty$.

For the case of a symmetric potential such as $V_2(x)$, the perturbative corrections to the energy arising from ΔV can be most simply calculated by use of the logarithmic perturbation theory. Logarithmic perturbation theory is the method of choice of doing perturbation theory for one dimensional potentials. A discussion of this method is found in the article of Imbo and Sukhatme and is summarized in Appendix C.

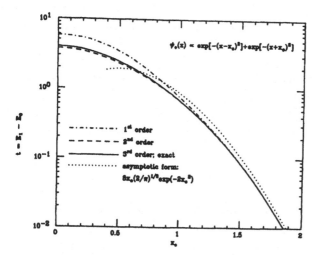

$$\psi_0(x) \propto \exp[-(x-x_0)^2] + \exp[-(x+x_0)^2]$$

1st order
2nd order
3rd order; exact
asymptotic form:
$8x_0(2/\pi)^{1/2}\exp(-2x_0^2)$

Fig. 9.3   The energy splitting $t = E_1 - E_0$ as a function of the separation $2x_0$ of the superposed Gaussians in the ground state wave function $\psi_0(x)$.

The first and second order corrections to the unperturbed energy $E = 0$ are

$$E^{(1)} = \frac{1}{2\xi(0)} \,, \quad \xi(x) \equiv \int_x^\infty \phi^2(x')dx \,,$$

$$E^{(2)} = -2 \int_0^\infty \left[\frac{E^{(1)}\xi(x')}{\phi(x')}\right]^2 dx' \,. \tag{9.43}$$

For our example, we numerically evaluate these corrections in order to obtain an estimate of t. The results are shown in Fig. 9.3 for values of $x_0 \leq 2$. Estimates of t correct to first, second, and third order calculated from logarithmic perturbation theory are compared with the exact result for V_2, obtained by the Runge-Kutta method. The asymptotic behavior of t given by eq. (9.39) is also shown. This asymptotic form can also be recovered from eq. (9.43) by a suitable approximation of the integrand in the large-x_0 limit. Even for values of $x_0 \leq 1/\sqrt{2}$, in which case $V_1(x)$ does not exhibit a double-well structure, the approximation technique is surprisingly good. The third-order perturbative result and the exact result are indistinguishable for all values of x_0.

In conclusion, we have demonstrated how SUSY can be used to calculate t, the energy splitting for a double well potential. Rather than calculating this splitting as a difference between the lowest-lying two states of $V_1(x)$, one can instead develop a perturbation series for the ground state energy t of the partner potential $V_2(x)$. By choosing as an unperturbed problem the potential whose solution is the normalizable zero-energy solution of $V_0(x)$, we obtain a very simple δ–function perturbation which produces a rapidly convergent series for t. The procedure is quite general and is applicable to any arbitrary double-well potential, including asymmetric ones. The numerical results are very accurate for both deep and shallow potentials.

9.4   Supersymmetry and the Large-N Expansion

The large-N method, where N is the number of spatial dimensions, is a powerful technique for analytically determining the eigenstates of the Schrödinger equation, even for potentials which have no small coupling constant and hence not amenable to treatment by standard perturbation theory. A slightly modified, physically motivated approach, called the "shifted large-N method" incorporates exactly known analytic results into the $1/N$ expansion, greatly enhancing its accuracy, simplicity and range of applicability. In this section, we will describe how the rate of convergence of shifted $1/N$ expansions can be still further improved by using the ideas of SUSY QM .

The basic idea of the $1/N$ expansion in quantum mechanics consists of solving the Schrödinger equation in N spatial dimensions, assuming N to be large, and taking $1/N$ as an "artificially created" expansion parameter for doing standard perturbation theory. At the end of the calculation, one sets $N = 3$ to get results for problems of physical interest in three dimensions.

For an arbitrary spherically symmetric potential $V(r)$ in N dimensions, the radial Schrödinger equation contains the effective potential

$$V_{\text{eff}}(r) = V(r) + \frac{(k-1)(k-3)\hbar^2}{8mr^2} \ , \quad k = N + 2l \ . \qquad (9.44)$$

It is important to note that N and l always appear together in the combination $k = N + 2l$. This means that the eigenstates, which could in principle have depended on the three quantities N, l, n, in fact only depend on k and n, where n is the radial quantum number which can take values $0,1,2,...$. One now makes a systematic expansion of eigenstates in the parameter $1/\overline{k}$, where $\overline{k} = k - a$. Of course, for very large values of N, the two choices $\bar{k}$ and k are equivalent. However, for $N = 3$ dimensions, a properly chosen shift a produces great improvement in accuracy and simplicity. At small values of r, the $n = 0$ wave function $\psi_0(r)$ has the behavior $r^{(k-1)/2}$. If one sets

$$\psi_0(r) = r^{(k-1)/2} \Phi_0(r) , \qquad (9.45)$$

where $\Phi_0(r)$ is finite at the origin, then eq. (9.45) readily gives the supersymmetric partner potential of $V_{\text{eff}}(r)$ to be

$$V_2(r) = V(r) + \frac{(k+1)(k-1)\hbar^2}{8mr^2} - \frac{\hbar^2}{m}\frac{d^2}{dr^2}ln\Phi_0(r) . \qquad (9.46)$$

Note that $V_2(r)$ and $V_{\text{eff}}(r)$ have the same energy eigenvalues [except for the ground state]. However, large-N expansion with the partner potential $V_2(r)$ is considerably better since the angular momentum barrier in eq. (9.46) is given by $(k' - 1)(k' - 3)\hbar^2/8mr^2$, where $k' = k + 2$. So, effectively, one is working in two extra spatial dimensions! Thus, for example, in order to calculate the energy of the state with quantum numbers k, n of $V_{\text{eff}}(r)$ one can equally well use $k' = k + 2, n - 1$ with $V_2(r)$. To demonstrate this procedure, let us give an explicit example. Using the usual choice of units $\hbar = 2m = 1$, the s-wave Hulthen effective potential in three dimensions and its ground state wave function are:

$$V_{\text{eff}}^H(r) = -\frac{2\delta e^{-\delta r}}{1 - e^{-\delta r}} + \frac{(2-\delta)^2}{4} , \quad \psi_0(r) \sim \left(1 - e^{-\delta r}\right) e^{-\frac{(2-\delta)r}{2}} , \qquad (9.47)$$

where the parameter δ is restricted to be less than 2. The supersymmetric partner potential turns out to be

$$V_2^H(r) = V_{\text{eff}}^H(r) + \frac{2\delta^2 e^{-\delta r}}{(1 - e^{-\delta r})^2} . \qquad (9.48)$$

As r tends to zero, V_2^H goes like $2r^{-2}$, which as mentioned above, (note $\hbar = 2m = 1$) corresponds to the angular momentum barrier $(k' - 1)(k' - 3)/4r^2$ with $k' = 5$ ($N = 5, l = 0$). Let us compute the energy of the first excited state of $V_{\text{eff}}^H(r)$. For the choice $\delta = 0.05$, the exact answer is known to be 0.748125. The results up to leading, second and third order using a shifted $1/N$ expansion for V_{eff}^H are 0.747713, 0.748127 and 0.748125 . The corresponding values using the supersymmetric partner potential V_2^H are all 0.748125 ! It is clear that although excellent results are obtained with the use of the shifted $1/N$ expansion for the original potential $V_{\text{eff}}^H(r)$ in three dimensions, even faster convergence is obtained by using the supersymmetric partner potential, since we are now effectively working in five dimensions instead of three. Thus, SUSY has played an important role in making a very good expansion even better. In fact, for many applications, considerable analytic simplification occurs since it is sufficient to just use the leading term in the shifted $1/N$ expansion for $V_2(r)$.

References

(1) C. Bender, K. Milton, M. Moshe, S. Pinsky, and L. Simmons, *Novel Perturbative Scheme In Quantum Field Theory*, Phys. Rev. **D37** (1988) 1472-1484.

(2) F. Cooper, J. Dawson and H. Shepard, *SUSY Inspired Variational Method for the Anharmonic Oscillator*, Phys. Lett. **A187** (1994) 140-144.

(3) F. Cooper and P. Roy, *Delta Expansion for the Superpotential*, Phys. Lett. **A143** (1990) 202-206.

(4) T. Imbo and U. Sukhatme, *Logarithmic Perturbation Expansions in Nonrelativistic Quantum Mechanics*, Am. Jour. Phys. **52** (1984) 140-146.

(5) W.-Y. Keung, E. Kovacs and U. Sukhatme, *Supersymmetry and Double Well Potentials*, Phys. Rev. Lett. **60** (1988) 41-44.

(6) L. Landau and E. Lifshitz, *Nonrelativistic Quantum Mechanics*, Pergamon Press (1977).

(7) E. Merzbacher, *Quantum Mechanics*, Wiley (1970).

(8) L. Mlodinow and N. Papanicolaou, *SO(2,1) Algebra and the Large N Expansion in Quantum Mechanics*, Ann. Phys. **128** (1980) 314-334; *Pseudospin Structure and Large N Expansion for a Class of Generalized Helium Hamiltonians*, ibid. **131** (1981) 1-35.

(9) U. Sukhatme and T. Imbo, *Shifted 1/N Expansions for Energy Eigenvalues of the Schrödinger Equation*, Phys. Rev. **D28** (1983) 418-420.

Problems

1. Consider the partner Hamiltonians $H_{2,1} = p^2/2 + V_{2,1}$, with
$V_{2,1} = \frac{1}{2}(W^2 \pm W'); W = x^3$.
(a) Determine the ground state wave function and energy for H_1 exactly.
(b) Using the trial wave function of the form given in eq. (9.9) determine the energy functional $E_0(\rho, n)$.
(c) For $n = 1, 2$ find the value of ρ that minimizes the energy functionals for both H_2 and H_1.
(d) Determine the ground state energy from the results of (c). What can you say about the accuracy of this approximation for $n = 1, 2$ for H_2 and H_1?

2. There are some other polynomial potentials for which the ground state is exactly known. One has that for

$$2V = x^6 - 7x^2 : E_0 = -\sqrt{2} ; \quad 2V = x^6 - 11x^2 : E_0 = -4 .$$

Again calculate the variational energies for $n = 1, 2$ and compare with these exact results.

3. An alternative to the δ expansion of the text is what is known as the linear δ expansion. Again consider the partner potential $V_2 = \frac{1}{2}(x^6 + 3x^2)$ to be written as

$$V(x) = \frac{1}{2}(x^6 - 3x^2 + 6\,\delta\,x^2) ,$$

where again δ is initially assumed to be a perturbation parameter which will be set to one or extrapolated to one at the end. Following the discussion of the δ expansion, or following the related discussion of logarithmic perturbation theory of Appendix C, introduce a new potential

$$2V_1(x) = W^2 - W' = x^6 - 3x^2 + 6\delta x^2 - 2E(\delta) .$$

Assume all quantities have a series expansion in δ. First rederive that

$W_0 = x^3, E_0 = 0$. Next show that W_1 obeys the differential equation (see also eq. (C.3)) :

$$W_1' - 2x^3 W_1 = 6x^2 - E_1 .$$

Assuming a solution of the form $W(x) = e^{x^4/2} f(x)$, show that:

$$W_1(x) e^{-x^4/2} = \int_0^x dy \, e^{-y^4/2} [6y^2 - 2E_1] .$$

Therefore derive that

$$E_1 = 3\sqrt{2} \frac{\Gamma(3/4)}{\Gamma(1/4)} = 1.433397 .$$

If we just set $\delta = 1$ this then gives a higher answer for the ground state energy than the variational method. This would be improved if we instead calculated several terms in the linear δ expansion and extrapolated to $\delta = 1$ using say Padé approximants.

4. This problem describes the procedure of obtaining a $1/N$ expansion for the spherically symmetric power law potentials $V(r) = Ar^\nu$. Start from the radial Schrödinger equation in N dimensions. Scale out the characteristic distance of the problem $r_c \equiv (\hbar^2/2mA)^{1/(\nu+2)}$ and the corresponding characteristic energy $E_c \equiv Ar_c^\nu$, by defining $\xi \equiv r/r_c$ and $\lambda \equiv E/E_c$. Show that the Schrödinger equation now reads

$$\left[-\frac{d^2}{d\xi^2} + \xi^\nu + \frac{(k-1)(k-3)}{4\xi^2} \right] \phi = \lambda \phi ,$$

where $k = N + 2l$. Change the independent variable to $\eta = \xi k^{-2/(\nu+2)}$ and show that the effective potential at large k is $V_{eff} = \eta^\nu + 1/4\eta^2$ with a minimum at $\eta_0 = (2\nu)^{-1/(\nu+2)}$. Again make a change of variable $x = k^{1/2}(\eta - \eta_0)/\eta_0$ and show that the Schrödinger equation becomes

$$\left[-\frac{d^2}{dx^2} + \frac{k}{4}\left(1 - \frac{1}{k}\right)\left(1 - \frac{3}{k}\right)\left(1 + \frac{x}{k^{1/2}}\right)^{-2} + \frac{k}{2\nu}\left(1 + \frac{x}{k^{1/2}}\right)^\nu \right] \phi = \bar{\lambda} \phi$$

where $\bar{\lambda} = \lambda \eta_0^2 k^{\frac{2-\nu}{2+\nu}}$. Expand all terms in $1/k$ and equate order by order to get the result

$$E_n = \frac{\hbar^2}{2m} \left(\frac{4\nu A m k^\nu}{\hbar^2} \right)^{\frac{2}{2+\nu}} \left[\frac{2+\nu}{4\nu} + \frac{1}{k} \left\{ \left(n + \frac{1}{2} \right) \sqrt{2+\nu} - 1 \right\} + O(k^{-2}) \right] .$$

5. The rate of convergence of the $1/k$ expansion, discussed in the previous problem for power law potentials, can be substantially improved by changing to a $1/\bar{k}$ expansion, which makes use of the shifted expansion parameter $\bar{k} = k - 2 + (2n + 1)\sqrt{2+\nu}$. Here, the shift has been chosen so that the eigenvalues are exact for the analytically known cases $\nu = -1$(Coulomb) and $\nu = 2$(harmonic oscillator). Show that in this case, if one carries out the perturbation expansion up to $O(\bar{k}^{-2})$, the result is

$$E_n = \frac{\hbar^2}{2m} \left(\frac{4\nu A m \bar{k}^\nu}{\hbar^2} \right)^{\frac{2}{2+\nu}} \left[\frac{2+\nu}{4\nu} + \frac{(1+\nu)(2-\nu)}{72\bar{k}^2}(1 + 6n + 6n^2) + \dots \right] .$$

Appendix A

Path Integrals and SUSY

A.1 Dirac Notation

Quantum mechanics describes the state of a particle by a state vector $|\phi\rangle$ which belongs to a Hilbert space $\mathcal{H}$. $\mathcal{H}$ is the vector space of complex, square integrable functions, defined in configuration space. The scalar product of vectors in the space $\mathcal{H}$ in Dirac notation is

$$\langle \phi | \psi \rangle = \int d^3 r \phi^*(\vec{r}) \psi(\vec{r}) . \tag{A.1}$$

By definition, a vector $|\phi\rangle$ belongs to the Hilbert space $\mathcal{H}$ if the norm of $|\phi\rangle$ is finite:

$$\langle \phi | \phi \rangle = \int d^3 r \phi^*(\vec{r}) \phi(\vec{r}) < \infty . \tag{A.2}$$

The eigenvectors of the position operator $\hat{x}_i$ and momentum operator $\hat{p}_i$

$$\hat{x}_i |x_i\rangle = x_i |x_i\rangle; \quad \hat{p}_i |p_i\rangle = p_i |p_i\rangle , \tag{A.3}$$

do not belong to $\mathcal{H}$ because their norm is infinite, however they obey the closure relations:

$$\int d^3 r |\vec{r}\rangle\langle\vec{r}| = \int d^3 p |\vec{p}\rangle\langle\vec{p}| = 1 . \tag{A.4}$$

The overlap of these eigenvectors is given by

$$\langle \vec{r} | \vec{r}' \rangle = \delta^3(\vec{r} - \vec{r}') ; \quad \langle \vec{p} | \vec{p}' \rangle = \delta^3(\vec{p} - \vec{p}') , \tag{A.5}$$

and

$$\langle \vec{r} | \vec{p} \rangle = (\frac{1}{2\pi\hbar})^3 e^{\frac{i}{\hbar}\vec{p}\cdot\vec{r}} .$$

The wave function of a particle in the state $|\phi\rangle$ is given in the coordinate representation by

$$\phi(\vec{r}) = \langle \vec{r} | \phi \rangle . \qquad (A.6)$$

In the coordinate representation one has

$$\langle \vec{r} | \hat{r}_i | \phi \rangle = r_i \langle \vec{r} | \phi \rangle ,$$
$$\langle \vec{r} | \hat{p}_i | \phi \rangle = \frac{\hbar}{i} \frac{\partial}{\partial r_i} \langle \vec{r} | \phi \rangle . \qquad (A.7)$$

In what follows we will use the notation x, p to represent $\vec{r}, \vec{p}$. So for example the closure relationship will be

$$\int dx |x\rangle \langle x| = 1 .$$

A.2   Path Integral for the Evolution Operator

The matrix element of the evolution operator

$$\mathcal{U}(x_f, t_f; x_i, t_i) \equiv \langle x_f, t_f | x_i, t_i \rangle = \langle x_f | e^{-\frac{i}{\hbar}\hat{H}(t_f - t_i)} | x_i \rangle , \qquad (A.8)$$

can be cast in the form of a Feynman path integral by dividing the finite time interval into a large number of small steps and evaluating the evolution operator for each step. Let us divide the time interval $t_f - t_i$ into $N + 1$ equal steps of size ϵ,

$$\epsilon = \frac{t_f - t_i}{N + 1} . \qquad (A.9)$$

Then the intermediate times are denoted by

$$t_n = t_i + n\epsilon; \quad t_0 = t_i; \quad t_{N+1} = t_f , \qquad (A.10)$$

and the intermediate positions by

$$x_0 = x_i \text{ and } x_{N+1} = x_f .$$

Inserting the resolution of the identity

$$\int dx_n |x_n\rangle\langle x_n| = 1 \; ,$$

N times we have

$$\langle x_f, t_f | x_i, t_i \rangle = \prod_{k=0}^{N} \int dx_k \prod_{n=1}^{N+1} \langle x_n, t_n | x_{n-1}, t_{n-1} \rangle \; . \tag{A.11}$$

The object that needs to be evaluated is the matrix element of the infinitesimal evolution operator:

$$\langle x_n | e^{-i\frac{\epsilon}{\hbar} \hat{H}(\hat{p}, \hat{x})} | x_{n-1} \rangle = \int dp_n \langle x_n | p_n \rangle \langle p_n | e^{-i\frac{\epsilon}{\hbar} \hat{H}(\hat{p}, \hat{x})} | x_{n-1} \rangle \; . \tag{A.12}$$

Because the operators $\hat{p}$ and $\hat{x}$ do not commute one needs to find an approximation to the infinitesimal evolution operator which allows simple evaluation of the matrix element in eq. (A.12). There are several choices here, many of which are discussed in the book by Schulman. A very convenient choice is found in the book of Negele and Orland. There they define the normal form : $O(\hat{p}\hat{x})$: of an operator O to have all the $\hat{p}$ to the left and all the $\hat{x}$ to the right so that

$$: e^{-i\frac{\epsilon}{\hbar} \hat{H}(\hat{p}, \hat{x})} := \sum_{n=0}^{\infty} (-i\frac{\epsilon}{\hbar})^n \sum_{k=0}^{n} \frac{1}{k!(n-k)!} (\frac{\hat{p}^2}{2m})^k ((V(\hat{x}))^{n-k} \; . \tag{A.13}$$

They then prove that the difference between the exact infinitesimal evolution operator and the normal ordered one is of order ϵ^2, with the leading correction being

$$-\frac{\epsilon^2}{2\hbar^2} \left[V, \frac{\hat{p}^2}{2m} \right] \; . \tag{A.14}$$

We therefore obtain using eq. (A.12) and the normal ordered approximation that

$$\langle x_n | e^{-i\frac{\epsilon}{\hbar} \hat{H}(\hat{p}, \hat{x})} | x_{n-1} \rangle = \tag{A.15}$$

$$\int [dp]_n \exp \left[\frac{i}{\hbar} p_n \cdot (x_n - x_{n-1}) - i\frac{\epsilon}{\hbar} (\frac{p_n^2}{2m} + V(x_{n-1})) \right] + O(\epsilon^2) \; ,$$

where

$$[dp] = (\frac{dp}{2\pi\hbar})^d \; ,$$

and d is the number of spatial dimensions of the problem (usually 1 or 3). The integration over p is a shifted Gaussian integral in d dimensions. Using the basic integration formula for a shifted Gaussian in one dimension:

$$\int_{-\infty}^{\infty} dy \; e^{-ay^2+by} = \sqrt{\frac{\pi}{a}} e^{b^2/4a}. \tag{A.16}$$

we obtain

$$\langle x_n | e^{-i\frac{\epsilon}{\hbar}\hat{H}(\hat{p},\hat{x})} | x_{n-1} \rangle =$$
$$(\frac{m}{2\pi i\epsilon\hbar})^{d/2} e^{\frac{i\epsilon}{\hbar}\left(\frac{m}{2\epsilon^2}(x_n-x_{n-1})^2 - V(x_{n-1})\right)} + O(\epsilon^2) . \tag{A.17}$$

Therefore the matrix element of the evolution operator is approximately given by

$$\langle x_f, t_f | x_i, t_i \rangle = lim_{N\to\infty} \prod_{k=0}^{N} \times \tag{A.18}$$

$$\int dx_k (\frac{m}{2\pi i\epsilon\hbar})^{d(N+1)/2} \exp\left[\frac{i\epsilon}{\hbar} \sum_{n=1}^{N+1} \left(\frac{m}{2\epsilon^2}(x_n - x_{n-1})^2 - V(x_{n-1})\right)\right] .$$

The set of points $\{x_0, x_1, \cdots, x_{N+1}\}$ defines a trajectory in the limit $N \to \infty$ which is denoted by $x(t)$ with $x(t_i) = x_i$ and $x(t_f) = x_f$. In the continuum limit one has

$$\epsilon \sum_{n=1}^{N+1} \frac{m}{2}(\frac{x_n - x_{n-1}}{\epsilon})^2 \to \int_{t_i}^{t_f} dt \frac{m}{2}(\frac{dx}{dt})^2 ,$$
$$\epsilon \sum_{n=1}^{N+1} V(x_{n-1}) \to \int_{t_i}^{t_f} dt V(x(t)) , \tag{A.19}$$

so that in the continuum limit we get the Feynman path integral:

$$\langle x_f, t_f | x_i, t_i \rangle = \int_{(x_i,t_i)}^{(x_f,t_f)} \mathcal{D}[x(t)] e^{\frac{i}{\hbar} S[x(t)]} , \tag{A.20}$$

where

$$\int_{(x_i,t_i)}^{(x_f,t_f)} \mathcal{D}[x(t)] \equiv \lim_{N\to\infty} (\frac{m}{2\pi i\epsilon\hbar})^{d(N+1)/2} \prod_{k=1}^{N} \int dx_k ,$$

and the action $S[x(t)]$ is given by

$$S[x(t)] = \int_{t_i}^{t_f} dt L[x(t), \dot{x}(t)] = \int_{t_i}^{t_f} dt \left[\frac{m}{2} (\frac{dx}{dt})^2 - V(x(t)) \right] . \quad (A.21)$$

In the Heisenberg picture we often want to calculate the ground state correlation functions or Green's functions such as:

$$G(t_1, t_2) = \langle 0|T(\hat{x}(t_1)\hat{x}(t_2))|0\rangle , \quad (A.22)$$

where the time ordered product is given by

$$T(\hat{x}(t)\hat{x}(t')) = \Theta(t - t')\hat{x}(t)\hat{x}(t') + \Theta(t' - t)\hat{x}(t')\hat{x}(t) . \quad (A.23)$$

If we now assume $t' > t_1 > t_2 > t$ we have (inserting the resolution of the identity twice) that

$$G(t_1, t_2) = \int dx \, dx' \langle 0|x', t'\rangle \int_{x,t}^{x',t'} \mathcal{D}[x(t)]x(t_1)x(t_2)e^{\frac{i}{\hbar}\int_t^{t'} dtL[x(t)]}\langle x, t|0\rangle ,$$
$$(A.24)$$

where

$$\langle x, t|0\rangle = \phi_0(x)e^{-iE_0 t} ,$$

is the ground state wave function. A careful argument involving analytic continuation in time then shows that we can project out these ground state wave functions by letting $t \to (1 - i\eta)\tilde{t}$ with η an infinitesimal and then taking the limit $\tilde{t} \to \infty$. One then obtains

$$G(t_1, t_2) = \frac{\int \mathcal{D}[x(t)]x(t_1)x(t_2)e^{\frac{i}{\hbar}\int_{\tilde{t}=-\infty}^{\tilde{t}=\infty} dtL[x(t)]}}{\langle x', \tilde{t} \to \infty|x, \tilde{t} \to -\infty\rangle} . \quad (A.25)$$

It is convenient to define the functional $Z[j]$

$$Z[j] = N \int \mathcal{D}[x(t)]e^{\frac{i}{\hbar}\int_{-\infty}^{\infty} dt\{L[x(t)]+j(t)x(t)\}} \quad (A.26)$$

where the limit discussed above is now assumed. N is chosen so that

$$Z[j = 0] = 1 .$$

In terms of $Z[j]$, we find for example that

$$G(t_1, t_2) = \frac{\delta}{\delta j(t_1)} \frac{\delta}{\delta j(t_2)} Z[j]|_{j=0} , \quad (A.27)$$

which is the reason that $Z[j]$ is called the generating functional for the correlation functions.

A.3   Path Integrals for Fermionic Degrees of Freedom

A.3.1   *Hilbert Space for Fermionic Oscillator*

Two fermionic particles (for example, those having 1/2 integer spin) cannot occupy the same quantum state as a result of the the Pauli exclusion principle. The simplest fermionic system is described by the Hilbert space

$$\mathcal{H} = |0\rangle, |1\rangle \ , \tag{A.28}$$

with the property that the vacuum is defined by

$$a|0\rangle = 0 \ , \tag{A.29}$$

and the single particle state by

$$|1\rangle = a^\dagger |0\rangle \ . \tag{A.30}$$

The fermion creation and annihilation operators $a^\dagger, a$ obey the anticommutation relations

$$\begin{aligned}
\{a, a^\dagger\} &= aa^\dagger + a^\dagger a = 1 \ , \\
\{a, a\} &= \{a^\dagger, a^\dagger\} = 0 \ .
\end{aligned} \tag{A.31}$$

The Pauli principle is encoded in the fact that

$$a^\dagger |1\rangle = a^\dagger a^\dagger |0\rangle = 0 \ . \tag{A.32}$$

The completeness of the Hilbert space $\mathcal{H}$ is the relationship:

$$|0\rangle\langle 0| + |1\rangle\langle 1| = \mathbf{1} \ . \tag{A.33}$$

Coherent states of oscillators are states which are eigenfunctions of the annihilation operator i.e.

$$a|\phi\rangle = \phi|\phi\rangle \ . \tag{A.34}$$

We see that because of the anticommutation relations satisfied by the a, that the eigenvalues of a i.e. ϕ cannot be ordinary numbers but must be anticommuting variables that turn out to be members of what is known as a Grassmann algebra. That is we must have

$$\{\phi, \phi\} = 0 , \quad \{\phi, a\} = 0 . \tag{A.35}$$

The variables of a Grassmann algebra are necessary to encode into the definition of the fermion path integral the Pauli exclusion principle. The coherent states for a single oscillator are easy to construct as:

$$|\phi\rangle = e^{-\phi a^\dagger}|0\rangle = |0\rangle - \phi|1\rangle . \tag{A.36}$$

Using the anticommutation relation between ϕ and a as given by eq. (A.35), it is easy to see that eq. (A.34) is indeed satisfied. A Grassmann algebra is defined by a set of generators denoted by $\{\xi_\alpha\}, \alpha = 1, \cdots, n$. The generators anticommute:

$$\{\xi_\alpha, \xi_\beta\} = 0 . \tag{A.37}$$

The basis of the algebra is made of all distinct products of the generators:

$$\{1, \xi_{\alpha_1}, \xi_{\alpha_1}\xi_{\alpha_2}, \cdots \xi_{\alpha_1}\xi_{\alpha_2} \cdots \xi_{\alpha_n}\} .$$

If we consider a Grassmann algebra with two generators ξ, ξ^* then the basis is

$$\{1, \xi, \xi^*, \xi^*\xi\} , \tag{A.38}$$

and the most general object that can be written in this basis is of the form:

$$A(\xi^*, \xi) = a_0 + a_1\xi + \bar{a}_1\xi^* + a_{12}\xi^*\xi . \tag{A.39}$$

In terms of real Grassmann variables ξ_1 and ξ_2,

$$\xi = \frac{\xi_1 + i\xi_2}{\sqrt{2}}, \quad \xi^* = \frac{\xi_1 - i\xi_2}{\sqrt{2}} .$$

Next we want to define differentiation and integration over Grassmann variables. We define the derivative of a Grassmann variable in a similar fashion to that of a complex derivative except that in order for the derivative operator $\frac{\partial}{\partial \xi}$ to act on ξ the variable ξ must be anticommuted through the expression for A until it is adjacent to $\frac{\partial}{\partial \xi}$. For example:

$$\frac{\partial}{\partial \xi}(\xi^*\xi) = -\frac{\partial}{\partial \xi}(\xi\xi^*) = -\xi^* . \tag{A.40}$$

One finds that

$$\frac{\partial}{\partial \xi^*} \frac{\partial}{\partial \xi} A(\xi^*, \xi) = -a_{12} = -\frac{\partial}{\partial \xi} \frac{\partial}{\partial \xi^*} A(\xi^*, \xi) . \tag{A.41}$$

Thus the derivatives with respect to the Grassmann variables also anticommute:

$$\{\frac{\partial}{\partial \xi^*} \frac{\partial}{\partial \xi}, \frac{\partial}{\partial \xi} \frac{\partial}{\partial \xi^*}\} = 0 . \tag{A.42}$$

Integration over Grassmann variables is defined in such a manner as to give a definition for Gaussian integration which gave the correct effective action for an electron in an external electromagnetic field. These rules were codified in the book by Berezin and we will try to motivate them here. The main concept that one wants to preserve is analogous to the statement for ordinary integrals that if one has a function $f(x)$ which vanishes at $\pm\infty$ then we have

$$\int_{-\infty}^{\infty} dx \frac{df(x)}{dx} = f(x)|_{-\infty}^{\infty} = 0 . \tag{A.43}$$

Since 1 is the derivative of ξ this implies

$$\int d\xi \, 1 = 0 . \tag{A.44}$$

The only nonvanishing integral is that of ξ which is a constant conventionally taken to be 1, i.e.

$$\int d\xi \, \xi = 1 . \tag{A.45}$$

In eq. (A.45) one needs to make sure that ξ is anticommuted to be next to the $d\xi$. We notice that integration gives the same result as differentiation. To remove any ambiguities one needs to perform innermost integrals first, so that the convention for two Grassmann variables θ, η is

$$\int d\theta \int d\eta \, \eta\theta = +1 . \tag{A.46}$$

We have in analogy with differentiation that

$$\int d\xi^* d\xi A(\xi^*, \xi) = -a_{12} = -\int d\xi d\xi^* A(\xi^*, \xi) . \tag{A.47}$$

Another way the rules given by eqs. (A.44) and (A.45) are often obtained is by demanding the analogue of translation invariance for $f(x)$

$$\int_{-\infty}^{\infty} dx f(x+a) = \int_{-\infty}^{\infty} dx f(x) . \tag{A.48}$$

Now consider two Grassmann variables θ, η and a function

$$f(\theta) = A + B\theta , \tag{A.49}$$

so that under the shift $\theta \to \theta + \eta$ we demand

$$\int d\theta (A + B\theta) = \int d\theta \left((A + B\eta) + B\theta\right) = B . \tag{A.50}$$

This then leads to the rules (A.44) and (A.45). We next turn to the question of evaluating a Gaussian integral over a complex Grassmann variable. We have

$$\int d\xi^* d\xi e^{-\xi^* b\xi} = \int d\xi^* d\xi (1 - \xi^* b\xi) = \int d\xi^* d\xi (1 + \xi\xi^* b) = b . \tag{A.51}$$

This is to be contrasted with the result for complex integration over all space:

$$\int dz^* dz e^{-bz^* z} = \frac{2\pi}{b} . \tag{A.52}$$

So we see here that for fermions the factor of b occurs in the numerator rather than the denominator. If we have an additional factor of $\xi^* \xi$ we obtain instead

$$\int d\xi^* d\xi \ \xi^* \xi e^{-\xi^* b\xi} = -1 = -\frac{1}{b} \cdot b . \tag{A.53}$$

which yields (apart from the minus sign!) the same factor of $\frac{1}{b}$ we would have obtained from ordinary Gaussian integration. In what follows we will need to perform a general Gaussian integral in higher dimensions. In order to perform this integral we must first show that an integral over complex Grassmann variables is invariant under unitary transformations. We will then be able to evaluate the Gaussian integral involving an Hermitian matrix B with eigenvalues b_i by diagonalizing the matrix using a unitary transformation. First consider n complex Grassmann variables ξ_i and a unitary matrix U. If $\xi_i' = U_{ij}\xi_j$ then

$$
\begin{aligned}
\prod_i \xi_i' &= \frac{1}{n!}\epsilon^{ij\cdots l}\xi_i'\xi_j'\cdots\xi_l' \\
&= \frac{1}{n!}\epsilon^{ij\cdots l}U_{ii'}\xi_{i'}U_{jj'}\xi_{j'}\cdots U_{ll'}\xi_{l'} \\
&= \frac{1}{n!}\epsilon^{ij\cdots l}U_{ii'}U_{jj'}\cdots U_{ll'}\epsilon^{i'j'\cdots l'}(\prod_i \xi_i) \\
&= (\det\ U)(\prod_i \xi_i) \ .
\end{aligned}
\tag{A.54}
$$

In a general integral of the form

$$
\left(\prod_i \int d\xi_i^* d\xi_i\right) f(\xi^*, \xi) \ ,
\tag{A.55}
$$

the only term in $f(\xi^*, \xi)$ that survives has exactly one factor of each ξ_i and ξ_i^*; and is thus proportional to

$$
(\prod_i \xi_i)(\prod_i \xi_i^*) \ .
$$

Under the unitary transformation $\xi \to U\xi$ this term acquires a factor of

$$
(\det\ U)\ (\det\ U)^* = 1 \ .
$$

Thus the integral is unchanged by the unitary transformation. So now if we have a general Gaussian integral involving an Hermitian matrix B with eigenvalues b_i we obtain

$$
\left(\prod_i \int d\xi_i^* d\xi_i\right) e^{-\xi_i^* B_{ij}\xi_j} = \left(\prod_i \int d\xi_i^* d\xi_i\right) e^{-\xi_i^* b_i \xi_i} = \prod_i b_i = \det B \ .
\tag{A.56}
$$

This is the equation we will need to study the path integral formulation of SUSY quantum mechanics. Similarly one can show that

$$
\left(\prod_i \int d\xi_i^* d\xi_i\right) \xi_k \xi_l^* \ e^{-\xi_i^* B_{ij}\xi_j} = \det B (B^{-1})_{kl} \ .
\tag{A.57}
$$

A.4 Path Integral Formulation of SUSY Quantum Mechanics

In this appendix, we will describe the Lagrangian formulation of SUSY QM and discuss three related path integrals: one for the generating functional of correlation functions, one for the Witten index - a topological quantity which determines whether SUSY is broken, and one for a related "classical" stochastic differential equation, namely the Langevin equation. We will also briefly discuss the superspace formalism for SUSY QM. Starting from the matrix SUSY Hamiltonian which is 1/2 of our previous H [eq. (3.74)] for convenience:

$$H = \frac{1}{2}p^2 + \frac{1}{2}W^2(x)I - \frac{1}{2}[\psi, \psi^\dagger]W'(x) ,$$

we obtain the Lagrangian

$$L = \frac{1}{2}\dot{x}^2 + i\psi^\dagger \partial_t \psi - \frac{1}{2}W^2(x) + \frac{1}{2}[\psi, \psi^\dagger]W'(x) . \qquad (A.58)$$

In the above, H and L are operators in the Hilbert space which is the product of square integrable wave functions times two component column vectors describing the spin degrees of freedom. For the path integral the classical Lagrangian is needed. This requires replacing the operators ψ and $\psi^\dagger$ which act on the spinorial part of the wave function by the anti-commuting Grassman variables ψ and ψ^*, as well as x_{op} by the c number coordinate x. It is most useful to consider the generating functional of correlation functions in Euclidean space. We rotate $t \to i\tau$ and obtain for the Euclidean path integral :

$$Z[j, \eta, \eta^*] = \int [dx][d\psi][d\psi^*] \exp\left[-S_E + \int_0^\tau d\tau \, (jx + \eta\psi^* + \eta^*\psi)\right] , \quad (A.59)$$

where

$$S_E = \int_0^\tau d\tau \left(\frac{1}{2}x_\tau^2 + \frac{1}{2}W^2(x) - \psi^*[\partial_\tau - W'(x)]\psi\right) ,$$

and ψ and ψ^* are elements of a Grassmann algebra:

$$\{\psi^*, \psi\} = \{\psi, \psi\} = \{\psi^*, \psi^*\} = 0 , \qquad (A.60)$$

and

$$x_\tau = \frac{dx}{d\tau} \ .$$

The Euclidean action is invariant under the following SUSY transformations which mix bosonic and fermionic degrees of freedom:

$$\delta x = \epsilon^* \psi + \psi^* \epsilon \ ,$$

$$\delta \psi^* = -\epsilon^* \left(\partial_\tau x + W(x) \right) \ ,$$

$$\delta \psi = -\epsilon \left(-\partial_\tau x + W(x) \right) \ , \tag{A.61}$$

where ϵ and ϵ^* are two infinitesimal anticommuting parameters. These transformations correspond to $N = 2$ supersymmetry. The path integral over the fermions can now be explicitly performed using a cutoff lattice which is periodic in the the coordinate x but antiperiodic in the fermionic degrees of freedom at $\tau = 0$ and $\tau = T$. Namely we evaluate the fermionic path integral:

$$\int [d\psi][d\psi^*] \exp\left[\int_0^T d\tau \psi^* (\partial_\tau - W'(x)) \psi \right] \ , \tag{A.62}$$

by calculating the determinant of the operator $[\partial_\tau - W'(x)]$ using eigenvectors which are antiperiodic. We have, following Gildener and Patrascioiu, that

$$\det[\partial_\tau - W'(x)] = \prod_m \lambda_m \ ,$$

where

$$[\partial_\tau - W'(x)]\psi_m = \lambda_m \psi_m \ ,$$

so that

$$\psi_m(\tau) = C_m \exp[\int_0^\tau d\tau' [\lambda_m + W'] \ . \tag{A.63}$$

Imposing the antiperiodic boundary conditions:

$$\psi_m(T) = -\psi_m(0) \ ,$$

yields:

$$\lambda_m = \frac{i(2m+1)\pi}{T} - \frac{1}{T} \int_0^T d\tau W'(x) .$$ (A.64)

Regulating the determinant by dividing by the determinant for the case where the potential is zero we obtain

$$\det[\frac{\partial_\tau - W'(x)}{\partial_\tau}] = \cosh \int_0^T d\tau \frac{W'(x)}{2} .$$ (A.65)

Rewriting the cosh as a sum of two exponentials we find, as expected that Z is the sum of the partition functions for the two pieces of the supersymmetric Hamiltonian when the external sources are zero:

$$\mathrm{Tr}\, e^{-H_1 T} + \mathrm{Tr}\, e^{-H_2 T} \equiv Z_- + Z_+ .$$ (A.66)

Note that when SUSY is unbroken, only the ground state of H_1 contributes as $T \to \infty$. We also have

$$Z_\pm = \int [dx] \exp[-S_E^\pm] ,$$ (A.67)

where

$$S_E^\pm = \int_0^T d\tau \left(\frac{x_\tau^2}{2} + \frac{W^2(x)}{2} \pm \frac{W'(x)}{2} \right) .$$

A related path integral is obtained for the noise averaged correlation functions coming from a classical stochastic equation, the Langevin equation. What we have in mind here is a classical dynamical system being impinged upon by random sources. These random sources are assumed to have the property of white noise in that they are statistical in nature with the distribution being Gaussian at any particular time. In the following, these classical kicks at time τ are described by the random variable $\eta(\tau)$. It has been observed that for the particular stochastic differential equation

$$\dot{x} = W(x(\tau)) + \eta(\tau) ,$$ (A.68)

where $\eta(\tau)$ is a random stirring force obeying Gaussian statistics, the correlation functions of x are exactly the same as the correlation functions obtained from the Euclidean quantum mechanics related to the Hamiltonian

H_1. To see this we note that Gaussian noise is described by a probability functional:

$$P[\eta] = N \exp[-\frac{1}{2} \int_0^T d\tau \frac{\eta^2(\tau)}{F_0}] \,, \qquad (A.69)$$

normalized so that:

$$\int D\eta P[\eta] = 1 \,,$$

$$\int D\eta P[\eta]\eta(\tau) = 0 \,,$$

$$\int D\eta P[\eta]\eta(\tau)\eta(\tau') = F_0 \delta(\tau - \tau') \,,$$

so that the quantity F_0 describes the strength of the noise correlation function. The correlation functions averaged over the noise are

$$< x(\tau_1)x(\tau_2)... >= \int D\eta P[\eta]x(\tau_1)x(\tau_2)... \,, \qquad (A.70)$$

where we have in mind first solving the Langevin equation explicitly for $x(\eta(\tau))$ and then averaging over the noise. To make things more concrete, we can discretize the time

$$\tau \to na \equiv \frac{n}{\epsilon}$$

so that the discretized Langevin equation for $W(x) = gx$ is just:

$$\epsilon(x_n - x_{n-1}) = gx_n + \eta_n \,, \qquad (A.71)$$

which leads the update equation:

$$x_n = \frac{\epsilon}{\epsilon - g} x_{n-1} + \eta_n \,. \qquad (A.72)$$

In the regime where $g >> \epsilon$ one can analytically determine x_n in terms of η_i as a power series in $\frac{\epsilon}{g}$. On the lattice the path integral over η becomes a product of ordinary integrals at the discrete values of time $\tau_n = na$:

$$P[\eta] \to \prod_n [e^{-a\eta_n^2/2}(\frac{a}{2\pi})^{1/2}] \,. \qquad (A.73)$$

Another way to calculate the correlation functions without explicitly solving for x as a function of η is to change variables in the functional integral from η to x:

$$< x(\tau_1)x(\tau_2)... > = \int D[x] \; P[\eta] \; \det | \frac{d\eta(\tau)}{dx(\tau')} | \; x(\tau_1)x(\tau_2)... \; . \qquad (A.74)$$

This involves calculating the functional determinant,

$$\det | \frac{d\eta(\tau)}{dx(\tau')} | \; , \qquad (A.75)$$

subject to the boundary condition that the Green's functions obey causality, so one has retarded boundary conditions. One has

$$\det | \frac{d\eta}{dx} | = \exp \int_0^T d\tau \; \mathrm{Tr} \ln \left([\frac{d}{d\tau} - W'(x(\tau))]\delta(\tau - \tau') \right) \; . \qquad (A.76)$$

When there are no interactions ($W(x) = 0$), the retarded boundary conditions on the stochastic equation yield for the free Green's function that satisfies

$$\frac{dG_0}{d\tau} = \delta(\tau - \tau')$$

the result

$$G_0(\tau - \tau') = \theta(\tau - \tau') \; . \qquad (A.77)$$

Expanding the determinant around the free result by rewriting the ln in the form $\ln G_0^{-1}(1 - G_0 W')$, one finds because of the retarded boundary conditions that only the first term in the expansion contributes so that

$$\det | \frac{d\eta(\tau)}{dx(\tau')} | = \exp[\frac{1}{2} \int_0^T d\tau W'(x)] \; . \qquad (A.78)$$

Choosing $F_0 = \hbar$ so that

$$\begin{aligned}
P[\eta] &= N \exp[-\frac{1}{2} \int_0^T d\tau \frac{\eta^2(\tau)}{F_0}] \\
&= N \exp[-\frac{1}{2} \int_0^T d\tau (\dot{x} + W(x))^2] \\
&= N \exp[-\frac{1}{2} \int_0^T d\tau (\dot{x}^2 + W^2(x))] \; , \qquad (A.79)
\end{aligned}$$

we find that the generating functional for the correlation functions is exactly the generating functional for the correlation functions for Euclidean quantum mechanics corresponding to the Hamiltonian H_1:

$$Z[j] = N \int D[x] \exp\left[-\frac{1}{2} \int_0^T d\tau \left(x_\tau^2 + W^2(x) - W'(x) - 2j(\tau)x(\tau) \right) \right] .$$
(A.80)

Thus we see that we can determine the correlation functions of x for the Hamiltonian H_1 by either evaluating the path integral or by solving the Langevin equation and averaging over Gaussian noise. An equation related to the Langevin equation is the Fokker-Planck equation, which defines the classical probability function P_c for the equal time correlation functions of H_1. Defining the noise average:

$$P_c(z) = < \delta(z - x(t)) >_\eta = \int D\eta \delta(z - x(t))P[\eta] , \qquad (A.81)$$

one obviously has:

$$\int_{-\infty}^{\infty} dz z^n P_c(z,t) = \int D\eta [x(t)]^n P[\eta] = < x^n > .$$

One can show that P_c obeys the Fokker-Planck equation:

$$\frac{\partial P}{\partial t} = \frac{1}{2} F_0 \frac{\partial^2 P}{\partial z^2} + \frac{\partial}{\partial z}[W(z)P(z,t))] . \qquad (A.82)$$

For an equilibrium distribution to exist at long times t one requires that

$$P(z,t) \to \hat{P}(z) ,$$

and

$$\int_{-\infty}^{\infty} \hat{P}(z)dz = 1 .$$

Setting $\frac{\partial P}{\partial t} = 0$ in the Fokker-Planck equation, we obtain

$$\hat{P}(z) = N^2 \, e^{-2 \int_0^z W(y)dy} = \psi_0(z)^2 . \qquad (A.83)$$

Thus at long times only the ground state wave function contributes (we are in Euclidean space) and the probability function is just the usual ground state wave function squared. We see from this that when SUSY is broken, one cannot define an equilibrium distribution for the classical stochastic system. A third path integral for SUSY QM is related to the Witten index. As we discussed before, one can introduce a "fermion" number operator via

$$n_F = \frac{1 - \sigma_3}{2} = \frac{1 - [\psi, \psi^\dagger]}{2} . \tag{A.84}$$

Thus Witten index Δ can be written as

$$\Delta \equiv (-1)^F = [\psi, \psi^\dagger] = \sigma_3 . \tag{A.85}$$

The Witten index needs to be regulated and the regulated index is defined as:

$$\Delta(\beta) = \text{Tr} \, (-1)^F e^{-\beta H} = \text{Tr} \, (e^{-\beta H_1} - e^{-\beta H_2}) . \tag{A.86}$$

In Chapter 3 we discussed that the Witten index was important for understanding non-perturbative breaking of SUSY. Here we will show that the Witten index can be obtained using the path integral representation of the generating functional of SUSY QM where the fermion determinant is now evaluated using periodic boundary conditions to incorporate the factor $(-1)^F$. It is easy to verify a posteriori that this is the case. Consider the path integral:

$$\Delta(\beta) = \int [dx][d\Psi][d\Psi^*] e^{\int_0^\beta L_E(x, \Psi, \Psi^*) d\tau} , \tag{A.87}$$

where

$$L_E = \frac{1}{2} x_\tau^2 + \frac{1}{2} W^2 - \Psi^* [\partial_\tau - W'(x)] \Psi .$$

To incorporate the $(-1)^F$ in the trace, one changes the boundary conditions for evaluating the fermion determinant at $\tau = 0, \beta$ to periodic ones:

$$x(0) = x(\beta) ; \quad \Psi(0) = \Psi(\beta) .$$

The path integral over the fermions can again be explicitly performed using a cutoff lattice which is periodic in the fermionic degrees of freedom at $\tau = 0$ and $\tau = \beta$. We now impose these boundary conditions on the determinant of the operator $[\partial_\tau - W'(x)]$ using eigenvectors which are periodic. We again have

$$\det[\partial_\tau - W'(x)] = \prod_m \lambda_m .$$

Imposing periodic boundary conditions:

$$\lambda_m = \frac{i(2m)\pi}{\beta} - \frac{1}{\beta}\int_0^\beta d\tau W'(x) . \tag{A.88}$$

Regulating the determinant by dividing by the determinant for the case where the potential is zero we obtain

$$\det\left[\frac{\partial_\tau - W'(x)}{\partial_\tau}\right] = \sinh \int_0^\beta d\tau \frac{W'(x)}{2} . \tag{A.89}$$

Again rewriting the sinh as a sum of two exponentials we find, as expected that we obtain the regulated Witten index:

$$\Delta(\beta) = Z_- - Z_+ = \text{Tr } e^{-\beta H_1} - \text{Tr } e^{-\beta H_2} . \tag{A.90}$$

A.5 Superspace Formulation of SUSY Quantum Mechanics

One can think of SUSY QM as a degenerate case of supersymmetric field theory in $d = 1$ in the superspace formalism of Salam and Strathdee. Superfields are defined on the space $(x_n; \theta_a)$ where x is the space coordinate and θ_a are anticommuting spinors. In the degenerate case of $d = 1$ the field is replaced by $x(t)$ so that the only coordinate is time. The anticommuting variables are θ and θ^* where

$$\{\theta, \theta^*\} = \{\theta, \theta\} = [\theta, t] = 0 .$$

Consider the following SUSY transformations:

$$t' = t - i(\theta^*\epsilon - \epsilon^*\theta) , \quad \theta' = \theta + \epsilon , \quad \theta^{*\prime} = \theta^* + \epsilon^* . \tag{A.91}$$

If we assume that the generator of finite SUSY transformations is

$$L = e^{i(\epsilon^* Q^* + Q\epsilon)} ,$$

then from

$$\delta A = i[\epsilon^* Q^* + Q\epsilon, A] , \tag{A.92}$$

we infer that the operators Q and Q^* are given by

$$Q = i\partial_\theta - \theta^*\partial_t , \quad Q^* = -i\partial_{\theta^*} - \theta\partial_t . \tag{A.93}$$

Now these charges obey the familiar SUSY QM algebra:

$$\{Q, Q^*\} = 2i\partial_t = 2H \ , \quad [H, Q] = 0 \ , \{Q, Q\} = 0 \ . \tag{A.94}$$

The Lagrangian in superspace is determined as follows. A superfield made up of x , θ, and θ^* can at most be a bilinear in the Grassmann variables:

$$\phi(x, \theta, \theta^*) = x(t) + i\theta\psi(t) - i\psi^*\theta^* + \theta\theta^* D(t) \ . \tag{A.95}$$

Under a SUSY transformation, the following derivatives are invariant:

$$D_\theta = \partial_\theta - i\theta^* \partial_t$$

or in component form:

$$D_\theta \phi = i\psi - \theta^* D - i\theta^* \dot{x} + \theta^* \theta \dot{\psi} \ , \tag{A.96}$$

and

$$D_{\theta^*} = \partial_{\theta^*} - i\theta \partial_t \ ,$$

or in component form:

$$[D_\theta \phi]^* = -i\psi^* - \theta D + i\theta \dot{x} + \theta^* \theta \dot{\psi}^* \ . \tag{A.97}$$

The most general invariant action is:

$$S = \int dt d\theta^* d\theta \left(\frac{1}{2} |D_\theta \phi|^2 - f(\phi) \right) \ . \tag{A.98}$$

Again the expansion in terms of the Grassmann variables causes a Taylor expansion of f to truncate at the second derivative level. Integrating over the Grassmann degrees of freedom using the usual path integral rules for Grassmann variables:

$$\int \theta d\theta = \int \theta^* d\theta^* = 1, \quad \int d\theta = \int d\theta^* = 0 \ ,$$

one obtains

$$S = \int dt \left(\frac{1}{2} \dot{x}^2 + \psi^* [i\partial_t - f''(x)]\psi + \frac{1}{2} D^2 + D f'(x) \right) \ . \tag{A.99}$$

Eliminating the constraint variable $D = -f'(x) = W(x)$ we obtain our previous result for the action (now in Minkowski space):

$$S = \int dt \left(\frac{1}{2} \dot{x}^2 + \psi^* [i\partial_t - W'(x)]\psi - \frac{1}{2} W^2 \right) \ . \tag{A.100}$$

References

(1) J.W. Negele and H. Orland, *Quantum Many Particle Systems*, Perseus Books (1988).

(2) L.S. Schulman, *Techniques and Applications of Path Integration*, John Wiley (1981).

(3) F.A. Berezin, *The Method of Second Quantization*, Academic Press (1966).

(4) M.E. Peskin and D.V. Schroeder, *An Introduction to Quantum Field Theory*, Addison Wesley (1995).

(5) G. Parisi and N. Sourlas, *Supersymmetric Field Theories and Stocha tic Differential Equations*, Nucl. Phys. **B206** (1982) 321-332.

(6) P. Salomonson and J. van Holten, *Fermionic Coordinates and Supersymmetry in Quantum Mechanics*, Nucl. Phys. **B196** (1982) 509-531.

(7) F. Cooper and B. Freedman, *Aspects of Supersymmetric Quantum Mechanics*, Ann. Phys. **146** (1983) 262-288.

(8) E. Gildner and A. Patrascioiu, *Effect of Fermions Upon Tunnelling in a One-Dimensional System*, Phys. Rev. **D16** (1977) 1802-1804.

(9) A. Salam and J. Strathdee, *On Superfields and Fermi-Bose Symmetry*, Phys. Rev. **D11** (1975) 1521-1535.

Appendix B

Operator Transforms – New Solvable Potentials from Old

In 1971, Natanzon wrote down (what he thought at that time to be) the most general solvable potentials i.e. for which the Schrödinger equation can be reduced to either the hypergeometric or confluent hypergeometric equation. It turns out that most of these potentials are not shape invariant. Further, for most of them, the energy eigenvalues and eigenfunctions are known implicitly rather than explicitly as in the shape invariant case. One might ask if one can obtain these solutions from the explicitly solvable shape invariant ones. One strategy for doing this is to start with a Schrödinger equation which is exactly solvable (for example one having a SIP) and to see what happens to this equation under a point canonical coordinate transformation (PCT). In order for the Schrödinger equation to be mapped into another Schrödinger equation, there are severe restrictions on the nature of the coordinate transformation. Coordinate transformations which satisfy these restrictions give rise to new solvable problems. When the relationship between coordinates is implicit, then the new solutions are only implicitly determined, while if the relationship is explicit then the newly found solvable potentials are also shape invariant. In a more specific special application of these ideas, Kostelecky et al. were able to relate, using an explicit coordinate transformation, the Coulomb problem in d dimensions with the d-dimensional harmonic oscillator. Other explicit applications of the coordinate transformation idea are found in the review article of Haymaker and Rau.

Let us see how this works. We start from the one-dimensional Schrödinger equation

$$\left\{ -\frac{d^2}{dx^2} + [V(x) - E_n] \right\} \psi_n(x) = 0 . \tag{B.1}$$

Consider the coordinate transformation from x to z defined by

$$f(z) = \frac{dz}{dx} , \tag{B.2}$$

so that

$$\frac{d}{dx} = f \frac{d}{dz} . \tag{B.3}$$

The first step in obtaining a new Schrödinger equation is to change coordinates and divide by f^2 so that we have:

$$\left\{ -\frac{d^2}{dz^2} - \frac{f'}{f} \frac{d}{dz} + \frac{[V - E_n]}{f^2} \right\} \psi_n = 0 . \tag{B.4}$$

To eliminate the first derivative term, one next rescales the wave function:

$$\psi = f^{-1/2} \bar{\psi} . \tag{B.5}$$

Adding a term $\epsilon_n \bar{\psi}$ to both sides of the equation yields

$$-\frac{d^2 \bar{\psi}_n}{dz^2} + [\bar{V}(E_n) + \epsilon_n] \bar{\psi}_n = \epsilon_n \bar{\psi}_n , \tag{B.6}$$

where

$$\bar{V}(E_n) = \frac{V - E_n}{f^2} - \left[\frac{f'^2}{4f^2} - \frac{f''}{2f} \right] . \tag{B.7}$$

In order for this to be a legitimate Schrödinger equation, the potential $\bar{V}(E_n) + \epsilon_n$ must be independent of n. This can be achieved if the quantity G defined by

$$G = V - E_n + \epsilon_n f^2 , \tag{B.8}$$

is independent of n. How can one satisfy this condition? One way is to have f^2 and G to have the same functional dependence on $x(z)$ as the original potential V. This further requires that in order for $\bar{V}$ to be independent of n, the parameters of V must change with n so that the wave function corresponding to the n'th energy level of the new Hamiltonian is related to a wave function of the old Hamiltonian with parameters which depend on n. This can be made clear by a simple example.

Let us consider an exactly solvable problem–the three dimensional harmonic oscillator in a given angular momentum state with angular momentum β. The reduced ground state wave function for that angular momentum is

$$\psi_0(r) = r^{\beta+1} e^{-\alpha r^2/2} , \qquad (B.9)$$

so that the superpotential is given by

$$W(r) = \alpha r - (\beta + 1)/r , \qquad (B.10)$$

and H_1 is given by

$$H_1 = -\frac{d^2}{dr^2} + \frac{\beta(\beta + 1)}{r^2} + \alpha^2 r^2 - 2\alpha(\beta + 3/2) . \qquad (B.11)$$

By our previous argument we must choose $f = \frac{dz}{dr}$ to be of the form

$$f^2 = (\frac{dz}{dr})^2 = \frac{A}{r^2} + Br^2 + C . \qquad (B.12)$$

The solution of this equation gives $z = z(r)$ which in general is not invertible so that one knows $r = r(z)$ only implicitly as discussed before. However, for special cases one has an invertible function. Let us, for simplicity, now choose

$$f = r , \qquad z = r^2/2 . \qquad (B.13)$$

As discussed earlier, the energy eigenvalues of the three dimensional harmonic oscillator are give by

$$E_n = 4\alpha n , \qquad (B.14)$$

so that the condition we want to satisfy is

$$V - 4\alpha n + \epsilon_n f^2 = G = \frac{D}{r^2} + Er^2 + F . \qquad (B.15)$$

Equating coefficients we obtain

$$
\begin{aligned}
D &= \beta(\beta + 1) , \\
E &= \epsilon_n + \alpha^2 = \gamma , \\
F &= -4\alpha n - 2\alpha(\beta + 3/2) = -2Ze^2 . \qquad (B.16)
\end{aligned}
$$

We see that for the quantities D, E, F to be independent of n one needs to have that α, which describes the strength of the oscillator potential, be dependent on n. Explicitly, solving the above three equations and choosing $\beta = 2l + 1/2$, we obtain the relations:

$$\alpha(l, n) = \frac{Ze^2}{2(l + n + 1)} \;, \quad \epsilon_n = \gamma - \frac{Z^2 e^4}{4(l + 1 + n)^2} \;. \quad (B.17)$$

We now choose $\gamma = \alpha^2(l, n = 0)$ so that the ground state energy is zero. These energy levels are those of the hydrogen atom. In fact, the new Hamiltonian written in terms of z is now

$$\bar{H} = -\frac{d^2}{dz^2} + \frac{l(l+1)}{z^2} - \frac{Ze^2}{z} + \frac{Z^2 e^4}{4(l+1)^2} \;, \quad (B.18)$$

and the ground state wave function of the hydrogen atom is obtained from the ground state wave function of the harmonic oscillator via

$$\bar{\psi}_0 = f^{1/2} \psi_0 = z^{l+1} e^{-\alpha(l, n=0)z} \;. \quad (B.19)$$

Higher wave functions will have values of α which depend on n so that the different wave functions correspond to harmonic oscillator solutions with different strengths. All the exactly solvable shape invariant potentials of Table 4.1 can be inter-related by point canonical coordinate transformations. This is nicely illustrated in Fig.B.1. In general, r cannot be explicitly found in terms of z, and one has

$$dz/dr = f = \sqrt{A/r^2 + Br^2 + C} \;, \quad (B.20)$$

whose solution is:

$$\begin{aligned}
z \;=\; & \frac{\sqrt{A + Cr^2 + Br^4}}{2} + \frac{\sqrt{A}\,\log(r^2)}{2} \\
& - \frac{\sqrt{A}\,\log(2A + Cr^2 + 2\sqrt{A}\sqrt{A + Cr^2 + Br^4})}{2} \\
& + \frac{C\,\log(C + 2Br^2 + 2\sqrt{B}\sqrt{A + Cr^2 + Br^4})}{4\sqrt{B}} \;. \quad (B.21)
\end{aligned}$$

This clearly is not invertible in general. If we choose this general coordinate transformation, then the potential that one obtains is the particular class of Natanzon potentials whose wave functions are confluent hypergeometric functions in the variable r and are thus only implicitly known in terms of the true coordinate z. In fact, even the expression for the transformed

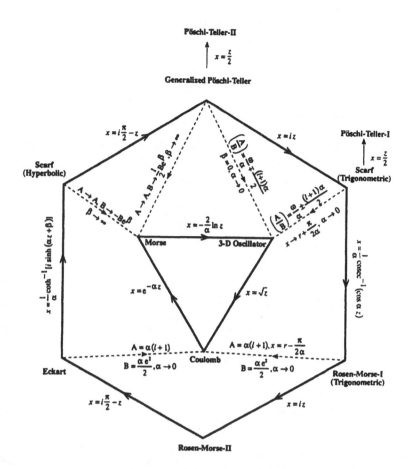

Fig. B.1 Diagram showing how all the shape invariant potentials of Table 4.1 are inter-related by point canonical coordinate transformations. Potentials on the outer hexagon have eigenfunctions which are hypergeometric functions whereas those on the inner triangle have eigenfunctions which are confluent hypergeometric functions.

potential is only known in terms of r:

$$\bar{V}(z, D, E) = 1/f^2[D/r^2 + Er^2 + F - f'^2/4 + ff''/2] , \qquad (B.22)$$

and thus only implicitly in terms of z. Equating coefficients, we get an implicit expression for the eigenvalues:

$$\frac{C\epsilon_n - F}{2\sqrt{E - B\epsilon_n}} - \sqrt{D + 1/4 - A\epsilon_n} = 2n + 1 , \qquad (B.23)$$

as well as the state dependence on α and β necessary for the new Hamiltonian to be energy independent:

$$\alpha_n = \sqrt{E - B\epsilon_n} \ , \qquad \beta_n = -1/2 + \sqrt{D + 1/4 - A\epsilon_n} \ . \qquad (B.24)$$

B.1   Natanzon Potentials

The more general class of Natanzon potentials whose wave functions are hypergeometric functions can be obtained by making an operator transformation of the generalized Pöschl-Teller potential whose Hamiltonian is

$$H_1 = A_1^\dagger A_1 = -\frac{d^2}{dr^2} + \frac{\beta(\beta - 1)}{\sinh^2 r} - \frac{\alpha(\alpha + 1)}{\cosh^2 r} + (\alpha - \beta)^2 \ . \qquad (B.25)$$

This corresponds to a superpotential

$$W = \alpha \tanh r - \beta \coth r \ , \quad \alpha > \beta \ , \qquad (B.26)$$

and a ground state wave function given by

$$\psi_0 = \sinh^\beta r \ \cosh^{-\alpha} r \ . \qquad (B.27)$$

The energy eigenvalues were discussed earlier and are

$$E_n = (\alpha - \beta)^2 - (\alpha - \beta - 2n)^2 \ . \qquad (B.28)$$

The most general transformation of coordinates from r to z which preserves the Schrödinger equation is described by

$$f^2 = \frac{B}{\sinh^2 r} - \frac{A}{\cosh^2 r} + C = (\frac{dz}{dr})^2 \ . \qquad (B.29)$$

From this we obtain an explicit expression for z in terms of r.

$$z =$$
$$\sqrt{A} \tan^{-1} \left[\frac{-3A + B - C + (A + B - C)\cosh 2r}{2G(r)\sqrt{A}} \right]$$
$$-\sqrt{B} \log \left[-A + 3B - C + (A + B + C)\cosh 2r \right.$$
$$\left. +2G(r)\sqrt{B} \right]$$
$$+\sqrt{C} \log \left[A + B + C\cosh 2r + G(r)\sqrt{C} \right]$$

$$+\sqrt{B}\,\log\left[2\sinh^2 r\right]\,, \tag{B.30}$$

where

$$G(r) = \sqrt{-2A + 2B - C + 2(A+B)\cosh 2r + C\cosh^2 2r}\,.$$

However the expression for the transformed potential is only known in terms of r:

$$\bar{V}(z,\gamma,\delta) = \frac{1}{f^2}\left[\frac{\delta(\delta-1)}{\sinh^2 r} - \frac{\gamma(\gamma+1)}{\cosh^2 r} + \sigma - \frac{f'^2}{4f^2} + \frac{f''}{2f}\right]\,, \tag{B.31}$$

and thus only implicitly in terms of z. Equating coefficients, we get an implicit expression for the eigenvalues:

$$[(\gamma+1/2)^2 - A\epsilon_n]^{1/2} - [(\delta-1/2)^2 - B\epsilon_n]^{1/2} - (\sigma - C\epsilon_n)^{1/2} = 2n+1\,, \tag{B.32}$$

as well as the state dependence on α and β necessary for the new Hamiltonian to be energy independent:

$$\begin{aligned}
\alpha_n &= [(\gamma+1/2)^2 - A\epsilon_n]^{1/2} - 1/2\,,\\
\beta_n &= [(\delta-1/2)^2 - B\epsilon_n]^{1/2} + 1/2\,.
\end{aligned} \tag{B.33}$$

Knowing the ground state wave function for the Natanzon potential we can determine the superpotential $W(r)$ and then find a hierarchy of Hamiltonians which generalize the Natanzon potential and whose wave functions are sums of hypergeometric functions. This class was first found by Cooper, Ginocchio and Khare in 1987.

References

(1) G.A. Natanzon, *Study of the One Dimensional Schrödinger Equation Generated from the Hypergeometric Equation*, Vestnik Leningrad Univ. **10** (1971) 22-28; *General Properties of Potentials for which the Schrödinger Equation Can be Solved by Means of Hypergeometric Functions*, Teoret. Mat. Fiz. **38** (1979) 219-229.

(2) V.A. Kostelecky, M.M. Nieto, D.Truax, *Supersymmetry and the Relationship Between the Coulomb and Oscillator Problems in Arbitrary Dimensions*, Phys. Rev. **D32** (1985) 2627-2651.

(3) R. Haymaker and A.R.P. Rau, *Supersymmetry in Quantum Mechanics*, Am. Jour. Phys. **54** (1986) 928-936.

(4) F. Cooper, J.N. Ginocchio and A. Khare, *Relationship Between Supersymmetry and Exactly Solvable Potentials*, Phys. Rev. **D36** (1987) 2458-2473.

(5) R. De, R. Dutt and U. Sukhatme, *Mapping of Shape Invariant Potentials Under Point Canonical Transformations*, Jour. Phys. **A25** (1992) L843-L850.

Appendix C

Logarithmic Perturbation Theory

Perturbation theory on bound states in nonrelativistic quantum mechanics is usually done using the Rayleigh-Schrödinger expansion in the coupling constant. While the calculation of the first order energy correction E_1 is straightforward, the formulas for higher order corrections E_n involve summations over all possible eigenstates which often cannot be explicitly performed even for simple perturbing potentials. Here, it is shown that if we restrict our attention to the important special case of perturbations to the ground state energy of one dimensional potentials, then the energy corrections to any order can be computed using an alternative much simpler form. This approach, called logarithmic perturbation theory (LPT) consists of first changing from the ground state wave function to the quantity $S(x) = \ln \psi(x)$, and then expanding in the coupling constant. LPT is considerably easier to use than other methods, since only a knowledge of the initial unperturbed ground state $\psi_0(x)$ is required and no summation over all intermediate states is necessary. More specifically, one does not need any information about the unperturbed excited states and the computations just involve well defined one dimensional integrals.

We now give the derivation of the LPT formulas. The time independent Schrödinger equation is

$$-\frac{\hbar^2}{2m}\frac{d^2\psi(x)}{dx^2} + [V_0(x) + gV_1(x)]\psi(x) = E\psi(x) , \qquad (C.1)$$

where $V_0(x)$ is the unperturbed potential and $V_1(x)$ is the perturbation with coupling constant g. After the transformation $\psi(x) = e^{S(x)}$, the

Schrödinger equation becomes

$$-\frac{\hbar^2}{2m}[S''(x) + S'(x)^2] + V_0(x) + gV_1(x) - E = 0 , \qquad \text{(C.2)}$$

which is Riccati's equation. Now expand E and $S'(x)$ in powers of the coupling constant g:

$$E = E_0 + gE_1 + g^2E_2 + \ldots ; \quad S'(x) = C_0(x) + gC_1(x) + g^2C_2(x) + \ldots .$$

Substitution into Riccati's equation, and collecting like powers of g yields the following series of equations:

$$C_0'(x) + C_0^2(x) = \frac{2m}{\hbar^2}[V_0(x) - E_0] , \qquad \text{(C.3)}$$

$$C_1'(x) + 2C_0(x)C_1(x) = \frac{2m}{\hbar^2}[V_1(x) - E_1] , \qquad \text{(C.4)}$$

$$C_n'(x) + 2C_0(x)C_n(x) = -\frac{2m}{\hbar^2}E_n - \sum_{s=1}^{n-1} C_s(x)C_{n-s}(x) , \; n = 2, 3, \ldots . \text{(C.5)}$$

Each of these linear differential equations is easily solved by using integration factors. Eq. (C.3) is just the unperturbed problem. Integration of eq. (C.4) using the integration factor $|\psi_0(x)|^2$ gives

$$\frac{d}{dx}[C_1(x)|\psi_0(x)|^2] = \frac{2m}{\hbar^2}[V_1(x) - E_1]|\psi_0(x)|^2 . \qquad \text{(C.6)}$$

Since $\psi_0(x) \to 0$ as $x \to -\infty$, we get

$$C_1(x)|\psi_0(x)|^2 = \frac{2m}{\hbar^2} \int_{-\infty}^{x} [V_1(x) - E_1]|\psi_0(x)|^2 \, dx . \qquad \text{(C.7)}$$

Since the left hand side goes to zero as $x \to \infty$, we get the standard result

$$E_1 = \int_{-\infty}^{\infty} V_1(x)|\psi_0(x)|^2 \, dx . \qquad \text{(C.8)}$$

Here we have assumed that the unperturbed ground state wave function $\psi_0(x)$ is normalized. Proceeding in the same manner, one can integrate eqs. (C.5) to get the following results:

$$E_2 = -\frac{\hbar^2}{2m} \int_{-\infty}^{\infty} C_1^2(x)|\psi_0(x)|^2 \, dx , \qquad \text{(C.9)}$$

$$C_2(x)|\psi_0(x)|^2 = -\int_{-\infty}^{x} \left[\frac{2m}{\hbar^2}E_2 + C_1^2(x)\right] |\psi_0(x)|^2 \, dx \, , \qquad \text{(C.10)}$$

$$E_3 = -\frac{\hbar^2}{2m} \int_{-\infty}^{\infty} 2C_1(x)C_2(x)|\psi_0(x)|^2 \, dx \, , \qquad \text{(C.11)}$$

which are readily generalized to arbitrary order n. Clearly the quantities $C_n(x)$ and E_n are determined by simple integration. Although we have kept absolute value signs for $|\psi_0(x)|^2$ in all the formulas these signs are superfluous since $\psi_0(x)$ can always be chosen to be real for one dimensional problems. Note that the quantities $C_n(x)$ are related to the perturbed ground state wave function and E_n give corrections to the ground state energy.

References

(1) Y. Aharonov and C. Au, *New Approach to Perturbation Theory*, Phys. Rev. Lett. **42** (1979) 1582-1585.
(2) T. Imbo and U. Sukhatme, *Logarithmic Perturbation Expansions in Nonrelativistic Quantum Mechanics*, Am. Jour. Phys. **52** (1984) 140-146.

Problems

1. Consider the situation of a charged particle in a harmonic oscillator potential $V_0(x) = \frac{1}{2}m\omega^2 x^2$. If an external constant electric field $\mathcal{E}$, producing a perturbing potential $V_1(x) = -e\mathcal{E}x$ is now turned on, compute the ground state energy using logarithmic perturbation theory. The calculation can be carried out to any order in the electric field $\mathcal{E}$. This problem is a simple variant of the Stark Effect.

2. The Coulomb potential for a hydrogen atom is $V_0(r) = -e^2/r$. The

ground state energy is $E_0 = -e^2/2a_0$, where $a_0 \equiv \hbar^2/me^2$ is the Bohr radius. A simple model which takes into account the effect of the nucleus is to consider the perturbation $V_1(r) = e^2/r - e^2/R$ for $r < R$, where R is the nuclear radius. Use logarithmic perturbation theory to show that the ground state energy is of the form

$$E = E_0[1 + \beta_2\eta^2 + \beta_3\eta^3 + O(\eta^4)] \; ,$$

where η denotes the dimensionless constant $\eta \equiv 2R/a_0$. Determine the numerical values of β_2 and β_3.

Appendix D

Solutions to Problems

Chapter 2.

Problem 2.1: Eigenfunctions correspond to an integer number of half wave-lengths fitted into the region $0 \le x \le L$. Normalized eigenfunctions and corresponding eigenenergies are

$$\psi_n(x) = \sqrt{\frac{2}{L}} \sin \frac{(n+1)\pi x}{L} \quad , \quad E_n = \frac{(n+1)^2 h^2}{8mL^2} \quad , \quad (n = 0, 1, 2, ...) \ .$$

Orthonormality follows from checking that $\int_0^L dx \psi_m(x)\psi_n(x) = \delta_{mn}$. Note that ψ_n has $(n+2)$ zeros located at

$$x = 0, L/(n+1), 2L/(n+1), ..., nL/(n+1), L \ ,$$

whereas ψ_{n+1} has $(n+3)$ zeros at

$$x = 0, L/(n+2), 2L/(n+2), ..., nL/(n+2), (n+1)L/(n+2), L \ .$$

Clearly ψ_{n+1} has exactly one zero located between the consecutive zeros of ψ_n.

Problem 2.2: (i) Taking the unit of energy to be $\hbar^2\pi^2/2ma^2$, for $V_0 = 0$ one has an infinite square well of width a with eigenenergies

$$E_n = (n+1)^2 \ ,$$

189

and for $V_0 = \infty$ one has an infinite square well of width $a/2$ with eigenenergies

$$E_n = 4(n+1)^2 \,, n = 0, 1, 2, \cdots \,.$$

(ii) By matching the wave function and its first derivative at $x = a/2$, one obtains a transcendental equation for the eigenenergies. The result depends on whether E is greater than or less than V_0:

$$\sqrt{E}\tan[\sqrt{(E-V_0)}\,\pi/2] + \sqrt{(E-V_0)}\tan[\sqrt{E}\,\pi/2] = 0 \,, (E > V_0) \,,$$

$$\sqrt{E}\tanh[\sqrt{(V_0-E)}\,\pi/2] + \sqrt{(V_0-E)}\tan[\sqrt{E}\,\pi/2] = 0 \,, (E < V_0) \,.$$

(iii) The values of the eigenenergies E_0 and E_1 for several choices of V_0 are tabulated below.

V_0	0	1	3	10	1000	∞
E_0	1.000	1.439	2.007	2.724	3.844	4.000
E_1	4.000	4.546	5.844	9.864	15.370	16.000

(iv) The critical value is $V_{0C} = 1.668$. For the special situation $E_0 = V_{0C}$, the ground state eigenfunction is linear in the region $a/2 \leq x \leq a$. The analytic expression for the un-normalized wave function is $\psi_0(x) = \sin kx$ for $0 \leq x \leq a/2$ and $\psi_0(x) = -k\cos(k\pi/2)(\pi - x)$ for $a/2 \leq x \leq a$, where $k \equiv \sqrt{V_{0C}} = 1.2916$.

Problem 2.3: Using units with $\hbar = 2m = 1$, the explicit low lying eigenfunctions for the harmonic oscillator potential $V(x) = \omega^2 x^2/4$ are given below, along with their zeros at finite values of x. Note that all bound state wave functions also vanish at $x = \pm\infty$.

n	$\psi_n(x)$	Zeros
0	$\exp(-\omega x^2/4)$	None
1	$x\exp(-\omega x^2/4)$	0
2	$(\omega x^2 - 1)\exp(-\omega x^2/4)$	$-\sqrt{1/\omega}, +\sqrt{1/\omega}$
3	$x(\omega x^2 - 3)\exp(-\omega x^2/4)$	$-\sqrt{3/\omega}, 0, +\sqrt{3/\omega}$

Problem 2.4: The Heisenberg equations of motion for x and p (using units

with $\hbar = 2m = 1$) are

$$\frac{dx}{dt} = 2p \ , \quad \frac{dp}{dt} = -\omega^2 x/2 \ .$$

The corresponding equations for a and $a^\dagger$ are

$$\frac{da}{dt} = -i\omega a \ , \quad \frac{da^\dagger}{dt} = i\omega a^\dagger \ .$$

Since the equations for a and $a^\dagger$ are decoupled, they are readily solved. The solutions are $a(t) = a(0)\exp(-i\omega t)$ and $a^\dagger(t) = a^\dagger(0)\exp(i\omega t)$. The unequal time commutators are

$$[x(t), x(t')] = -\frac{2i}{\omega}\sin[\omega(t - t')] \ ,$$

$$[p(t), p(t')] = -\frac{i\omega}{2}\sin[\omega(t - t')] \ ,$$

$$[x(t), p(t')] = i\cos[\omega(t - t')] \ .$$

Problem 2.5: The number operator is $N = a^\dagger a$. Since $N^2 = N$, the eigenvalues λ satisfy $\lambda^2 = \lambda$, which gives two solutions $\lambda = 0, 1$. Using the Fermi oscillator anti-commutation relations, the Hamiltonian can be re-written as $H = (-N + 1/2)\hbar\omega$, and the eigenvalues of H are $\pm\hbar\omega/2$.

Chapter 3.

Problem 3.1: An infinite square well of width π has been discussed in the text. The potential is $V_1(x) = 0 \ (0 < x < \pi)$, with eigenstates $E_n^{(1)} = (n + 1)^2$, $\psi_n^{(1)} \propto \sin(n + 1)x$, $n = 0, 1, 2, \ldots$. The superpotential is $W_1(x) = -\psi_0^{(1)'}(x)/\psi_0^{(1)}(x) = -\cot x$. The SUSY partner potential of $V_1(x)$ is $V_2(x) = V_1(x) + 2W_1'(x) = 2\mathrm{cosec}^2 x$, with eigenvalues $E_n^{(2)} = (n + 2)^2$, $n = 0, 1, 2, \ldots$, and eigenfunctions

$$\psi_n^{(2)} \propto A_1\psi_{n+1}^{(1)} = (d/dx - \cot x)\sin(n + 1)x \ .$$

The low lying eigenfunctions have the explicit form $\psi_0^{(2)} \propto \sin^2 x$, $\psi_1^{(2)} \propto \sin^2 x \cos x$, $\psi_2^{(2)} \propto \sin^2 x(6\sin^2 x - 5)$.

We use information about $V_2(x)$ to generate the next potential in the hierarchy. The superpotential is

$$W_2(\dot{x}) = -\psi_0^{(2)'}(x)/\psi_0^{(2)}(x) = -2\cot x \ .$$

The supersymmetric partner potential of $V_2(x)$ is $V_3(x) = V_2(x) + 2W_2'(x) = 6\,\mathrm{cosec}^2 x$, with eigenvalues $E_n^{(3)} = (n+3)^2$, $n = 0, 1, 2, \ldots$ and eigenfunctions $\psi_n^{(3)} \propto A_2\psi_{n+1}^{(2)} = (d/dx - 2\cot x)\psi_{n+1}^{(2)}$. The low lying eigenfunctions have the explicit form $\psi_0^{(3)} \propto \sin^3 x$, $\psi_1^{(3)} \propto \sin^3 x \cos x$.

Now, we use information about $V_3(x)$ to obtain the next potential in the hierarchy. The superpotential is $W_3(x) = -\psi_0^{(3)'}(x)/\psi_0^{(3)}(x) = -3\cot x$. The supersymmetric partner potential of $V_3(x)$ is $V_4(x) = V_3(x) + 2W_3'(x) = 12\,\mathrm{cosec}^2 x$, with eigenvalues $E_n^{(4)} = (n+4)^2$, $n = 0, 1, 2, \ldots$ and ground state eigenfunction $\psi_0^{(4)} = (d/dx - 3\cot x)\sin^3 x \cos x \propto \sin^4 x$.

The pattern is now clear. The general potential in the hierarchy is $V_m(x) = m(m-1)\mathrm{cosec}^2 x$ with ground state eigenfunction $\psi_0^{(m)} \propto \sin^m x$, and energy eigenvalues $E_n^{(m)} = (n+m)^2$, $n = 0, 1, 2, \ldots$. Similarly, the first excited state has the form $\psi_1^{(m)} \propto \sin^m x \cos x$. These results can of course be checked by direct substitution into the Schrödinger equation.

Problem 3.2: The supersymmetric partner potentials are $V_{2,1} = W^2 \pm W' = a^2x^6 \pm 3ax^2$. Both V_2 and V_1 are symmetric potentials. At large $\pm x$, they both have the same asymptotic behavior a^2x^6, whereas at small x, $V_2 \to +3ax^2$ and $V_1 \to -3ax^2$. $V_2(x)$ is a single well potential with a minimum at $x = 0$, whereas $V_1(x)$ is a double well potential with a local maximum at $x = 0$, and minima at $x = \pm a^{-1/4}$.

Problem 3.3: The superpotential is $W(r) = \beta - 5/r$, and the supersymmetric partner potentials are $V_1 = \beta^2 - 10\beta/r + 20/r^2$ and $V_2 = \beta^2 - 10\beta/r + 30/r^2$.

Problem 3.4: Since the quantity $e^{-\int W dx} = e^{-(Ax^3/3 + Bx^2/2 + Cx)}$ is not normalizable, this is an example of broken supersymmetry, and both V_2 and V_1 have identical energy spectra. The partner potentials $V_{2,1}(x) = W^2 \pm W' = (Ax^2 + Bx + C)^2 \pm (2Ax + B)$ are related to each other by a translation and reflection. A little algebra shows that $V_2(x - B/2A) = V_1(-x + B/2A)$. For the values $A = 1/5$, $B = 1$, $C = 0$, the partner potentials are $V_2(x) = x^4/25 + 2x^3/5 + x^2 + 2x/5 + 1$ and $V_1(x) = x^4/25 + 2x^3/5 + x^2 - 2x/5 - 1$. Note that V_1 and V_2 are also related by $V_2(x, A, B, C) = V_1(x, -A, -B, -C)$.

Problem 3.5: From the Rosen-Morse II entry in the table of shape invariant potentials given in the text, one sees that the potential $V_1(x) = -12 \operatorname{sech}^2 x$ has energy levels at $E_0^{(1)} = -9$, $E_1^{(1)} = -4$, $E_2^{(1)} = -1$, with an energy continuum for $E \geq 0$. The eigenfunctions are $\psi_n^{(1)} \propto \operatorname{sech}^{3-n} x P_n^{(3-n,3-n)}(\tanh x)$, $n = 0, 1, 2$. From the ground state wave function, the superpotential is found to be $W_1 = 3 \tanh x$. The partner potential of $V_1(x)$ is $V_2(x) = V_1(x) + 2W_1' = -6 \operatorname{sech}^2 x$.

It is easily checked that $V_2(x) = -6 \operatorname{sech}^2 x$ has energy levels at $E_0^{(2)} = -4$, $E_1^{(2)} = -1$, with an energy continuum for $E \geq 0$. The corresponding eigenfunctions are $\psi_0^{(2)}(x) \propto \operatorname{sech}^2 x, \psi_1^{(2)}(x) \propto \operatorname{sech} x \tanh x$. The ground state wave function $\psi_0^{(2)}$ yields the superpotential $W_2 = 2 \tanh x$. Repeating the same process again, gives $V_3(x) = -2\operatorname{sech}^2 x$ and $V_4(x) = 0$. It is easily checked that V_3 has only one bound state at $E_0^{(3)} = -1$ with the corresponding eigenfunction $\psi_0^{(3)}(x) \propto \operatorname{sech} x$ while V_4 obviously does not hold any bound state! As discussed in the text, V_1, V_2, V_3, V_4 are all reflectionless potentials so that the transmission coefficient is a pure phase. Using the Transmission coefficient $T_1(k)$ for the potential V_1 as given in chapter 2 and the relation (3.32) it is easily shown that

$$T_2(k) = \frac{\Gamma(-2 - ik)\Gamma(3 - ik)}{\Gamma(-ik)\Gamma(1 - ik)} \, ,$$

$$T_3(k) = \frac{\Gamma(-1 - ik)\Gamma(2 - ik)}{\Gamma(-ik)\Gamma(1 - ik)} \, ,$$

while obviously $T_4(k) = 1$.

Chapter 4.

Problem 4.1: The ground state wave function $\psi_0(r)$ has a maximum at $r = (1.5)^{1/2} r_0$. The superpotential and partner potentials are

$$W(r) = \frac{2r}{r_0^2} - \frac{3}{r} , \quad V_1(r) = \frac{4r^2}{r_0^4} - \frac{14}{r_0^2} + \frac{6}{r^2} , \quad V_2(r) = \frac{4r^2}{r_0^4} - \frac{10}{r_0^2} + \frac{12}{r^2} .$$

$V_1(r)$ has a minimum at $r = (1.5)^{1/4} r_0$. $V_2(r)$ has a minimum at $r = 3^{1/4} r_0$. From the superpotential and the table of exactly solvable shape invariant problems given in the text, one recognizes that this problem corresponds to a three dimensional oscillator with $\omega = 4/r_0^2$, $l = 2$. $V_1(r)$ has energy levels $E_n^{(1)} = \frac{8n}{r_0^2}$, $(n = 0, 1, 2, \ldots)$ and eigenfunctions $\psi_n^{(1)} \propto r^3 e^{-r^2/r_0^2} L_n^{5/2}(2r^2/r_0^2)$, where $L_n^{5/2}$ is a Laguerre polynomial. $V_2(r)$ has the same energy levels as $V_1(r)$, except that there is no zero energy eigenstate, since this is an example of unbroken supersymmetry. More precisely, the ground, first and second excited states of $V_1(r)$ are

$$E_0^{(1)} = 0 \quad ; \quad \psi_0^{(1)}(r) \propto r^3 e^{-r^2/r_0^2} ,$$
$$E_1^{(1)} = 8/r_0^2 \quad ; \quad \psi_1^{(1)}(r) \propto r^3 e^{-r^2/r_0^2}(4r^2 - 7r_0^2) ,$$
$$E_2^{(1)} = 16/r_0^2 \quad ; \quad \psi_2^{(1)}(r) \propto r^3 e^{-r^2/r_0^2}(16r^4 - 72r^2 r_0^2 + 63r_0^4) .$$

Similarly, the ground and first excited states of $V_2(r)$ are

$$E_0^{(2)} = 8/r_0^2 \quad ; \quad \psi_0^{(2)}(r) \propto r^4 e^{-r^2/r_0^2} ,$$
$$E_1^{(2)} = 16/r_0^2 \quad ; \quad \psi_1^{(2)}(r) \propto r^4 e^{-r^2/r_0^2}(4r^2 - 9r_0^2) .$$

Problem 4.2: There are three types of symmetric shape invariant potentials. The superpotential, potential and first three eigenvalues are:

$W(x)$	Potential $V_-(x)$	Eigenvalues E_0, E_1, E_2
$\omega x/2$	$\omega^2 x^2/4 - \omega/2$	$0, \omega, 2\omega$
$A \tanh \alpha x$	$A^2 - A(A+\alpha)\mathrm{sech}^2 \alpha x$	$0, A^2 - (A - \alpha)^2, A^2 - (A - 2\alpha)^2$
$A \tan \alpha x$	$-A^2 + A(A - \alpha)\sec^2 \alpha x$	$0, -A^2 + (A + \alpha)^2, -A^2 + (A + 2\alpha)^2$

Problem 4.3: The Hulthén potential is always attractive, diverges like $-aV_0/r$ at small r and goes exponentially to zero at large r. It can be re-written as

$$V(r) = -V_0 \frac{e^{-r/2a}}{[e^{r/2a} - e^{-r/2a}]} = -\frac{V_0}{2}[1 - \coth(r/2a)] \ ,$$

which is now recognizable as a special case of the Eckart potential in our list of analytically solvable potentials. The energy eigenvalues are $E_n = -[V_0 a^2 - (n+1)^2]^2/[4a^2(n+1)^2] \ , \ n = 0, 1, \ldots, n_{MAX}$. Here n_{MAX} is the largest positive integer such that $V_0 a^2 - (n+1)^2 > 0$.

Chapter 5.

Problem 5.1: It is easily seen that in this case the Pauli equation (for $\sigma = \pm 1$) reduces to the Schrödinger equations for the SUSY partner potentials

$$V_1(y) = A(A - \alpha) \sec^2 \alpha y + 2B \tan \alpha y + B^2/A^2 - A^2 \ ,$$

$$V_2(y) = A(A + \alpha) \sec^2 \alpha y + 2B \tan \alpha y + B^2/A^2 - A^2 \ ,$$

where $B/A = k + c$ and $-\pi/2 \leq \alpha y \leq \pi/2$. On comparing with Table 4.1 it is clear that these are shape invariant Rosen-Morse I potentials and hence the spectrum of $V_1(y)$ is

$$E_n = (A + n\alpha)^2 - A^2 + B^2/A^2 - B^2/(A + n\alpha)^2 \ , \ n = 0, 1, \ldots \ .$$

Problem 5.2: For the Lorentz scalar potential $\phi(x) = a\epsilon(x)$, the uncoupled equation for the Dirac spinor ψ_1 takes the form

$$\left[-\frac{d^2}{dx^2} + a^2 - 2a\delta(x) \right] \psi_1(x) = \omega^2 \psi_1(x) \ .$$

This is essentially the Schrödinger equation for an attractive delta function potential which is well known to have only one bound state at $\omega^2 = 0$ with the corresponding eigenfunction being

$$\psi_1(x) = e^{-ax\epsilon(x)} \ .$$

Hence the corresponding solution is

$$\psi(x) = \begin{pmatrix} \psi_1(x) \\ 0 \end{pmatrix}.$$

Problem 5.3: In this case, notice that the Lorentz scalar potential is antisymmetric over a half-period i.e. it satisfies

$$\phi(x + K(x)) = -\phi(x) \ .$$

Hence the corresponding partner potentials $V_{1,2}(x) = \phi^2(x) \mp \phi'(x)$ are self-isospectral. Now the uncoupled equation for $\psi_1(x)$ is

$$\left[-\frac{d^2}{dx^2} + 2m \ \text{sn}^2 x - m \right] \psi_1(x) = \omega^2 \psi_1(x) \ .$$

From Chapter 7 we know that in this case there is one bound band with band edges at $\omega^2 = 0, 1 - m$ while the continuum band begins at $\omega^2 = 1$. The corresponding band edge eigenfunctions respectively are

$$\text{dn} \, x, \ \text{cn} \, x, \ \text{sn} \, x \ .$$

Hence the solution of the Dirac problem is immediately written down:

$$\psi(x) = \begin{pmatrix} \psi_1(x) \\ \psi_1(x + K(x)) \end{pmatrix} \ .$$

Problem 5.4: If one has both scalar and vector potentials $U(r)$ and $V(r)$ respectively, then the Dirac Hamiltonian has the form

$$H = \alpha.\mathbf{p} + \beta(m + U) + V.$$

To compute the energy levels, one only needs to concentrate on the radial equations which now take the form

$$G'(r) + \frac{kG}{r} - [E + m + U - V]F = 0 \ ,$$

$$F'(r) - \frac{kF}{r} - [m - E + U + V]G = 0 .$$

Clearly, if $U(r) = \pm V(r)$ then one of the equation takes a simpler form and one can obtain an uncoupled second order equation for F or G. For example, consider $U(r) = V(r) = \omega^2 r^2/2$ in which case G satisfies the second order equation

$$\left[\frac{d^2}{dr^2} + E^2 - E\omega^2 r^2 - \frac{k(k+1)}{r^2} \right] G(r) = 0 .$$

The corresponding eigenvalues and eigenfunctions $G(r)$ are

$$E = [\omega(4n + 2k + 3)]^{2/3} , \quad n = 0, 1, ...$$

$$G(r) = r^{k+1} e^{-\omega\sqrt{E}r^2/2} L_n^{k+3/2}(\omega\sqrt{E}r^2) .$$

The corresponding function $F(r)$ is easily obtained.

Chapter 6.

Problem 6.1: Consider an infinite square well $V_1(x) = 0, 0 \le x \le L$. The normalized eigenfunctions are $\psi_n = \sqrt{\frac{2}{L}} \sin(\frac{(n+1)\pi x}{L})$, $n = 0, 1, 2, 3, ...$. Using the ground state wave function, the superpotential is

$$W(x) = -\psi_0'(x)/\psi_0(x) = -\frac{\pi}{L} \cot \frac{\pi x}{L} .$$

The one parameter family of isospectral potentials is

$$\hat{V}_1(x, \lambda) = -2\frac{d^2}{dx^2} \ln[I(x) + \lambda] , \quad 0 \le x \le L ,$$

where

$$I(x) \equiv \frac{2}{L} \int_0^x \sin^2 \frac{\pi x}{L} dx = \frac{x}{L} - \frac{1}{2\pi} \sin \frac{2\pi x}{L} .$$

The ground state of the potential $\hat{V}_1(x, \lambda)$ is

$$\hat{\psi}_0(x, \lambda) \propto \frac{\sin \frac{\pi x}{L}}{\frac{x}{L} - \frac{1}{2\pi} \sin \frac{2\pi x}{L} + \lambda} .$$

The excited states of $\hat{V}_1(x, \lambda)$ are readily computed from the expression

$$\hat{\psi}_n(x, \lambda) = \hat{A}^{\dagger}(\lambda) A \psi_n(x) = \left[-\frac{d}{dx} + \hat{W}(x, \lambda) \right] \left[\frac{d}{dx} + W(x) \right] \psi_n(x) ,$$

where,

$$\hat{W}(x, \lambda) = W(x) + \frac{(2/L) \sin^2 \frac{\pi x}{L}}{\frac{x}{L} - \frac{1}{2\pi} \sin \frac{2\pi x}{L} + \lambda} .$$

Problem 6.2: The normalized ground state wave function corresponding to the potential $V_1(x) = 1 - 2 \operatorname{sech}^2 x$ is $\psi_1(x) = \frac{1}{\sqrt{2}} \operatorname{sech} x$. The indefinite integral $I(x)$ is $(1 + \tanh x)/2$, and the one parameter family of potentials is $V_1(x, \lambda) = V_1(x) - 2 \frac{d^2}{dx^2} \ln(I(x) + \lambda)$. After substantial algebraic simplification, one gets the explicit simplified form $V_1(x, \lambda) = 1 - 2 \operatorname{sech}^2(x + a)$, where the quantity a is given by $\coth a = 1 + 2\lambda$, which is equivalent to $a = \frac{1}{2} \ln(1 + \lambda^{-1})$, as given in the text.

Problem 6.3: Since $V_1(x)$ is symmetric, so is its normalized ground state wave function $\psi_1(x)$. For a symmetric $\psi_1(x)$, it follows that $I(x) + I(-x) = 1$, where $I(x) \equiv \int_{-\infty}^{x} dx' \psi_1^2(x')$. The potential $V_1(x, \lambda) = V_1(x) - 2 \frac{d^2}{dx^2} [\ln\{I(x) + \lambda\}]$ is a member of the isospectral family of $V_1(x)$. The parity reflected potential is

$$\begin{aligned} V_1(-x, \lambda) &= V_1(-x) - 2 \frac{d^2}{dx^2} [\ln\{I(-x) + \lambda\}] \\ &= V_1(x) - 2 \frac{d^2}{dx^2} [\ln\{I(x) - \lambda - 1\}] = V_1(x, -\lambda - 1) . \end{aligned}$$

Clearly, $V_1(-x, \lambda)$ is also a member of the isospectral family of $V_1(x)$ corresponding to a parameter value $-\lambda - 1$.

Problem 6.4: The potential $V_1(x) = -6 \operatorname{sech}^2 x$ has two eigenstates at energies $E_1 = -4$, $E_2 = -1$. The normalized ground state wave function is $\psi_1(x) = \frac{\sqrt{3}}{2} \operatorname{sech}^2 x$, which yields the indefinite integral $I(x) = (2 + 3 \tanh x - \tanh^3 x)/4$. The one parameter family $V_1(x, \lambda)$ corresponds to putting $\lambda_1 = \lambda$ and $\lambda_2 = \infty$ ($\delta_2 = 0$), in the two parameter family $V_1(x, \lambda_1, \lambda_2)$ formula given in the text. The un-normalized ground state of $V_1(x, \lambda)$ is

$$\psi_1(x, \lambda) \propto \frac{\text{sech}^2 x}{2 + 4\lambda - 3\tanh^2 x - \tanh^3 x} .$$

Problem 6.5: Consider traveling wave solutions $u(x, t) = f(x - ct)$ moving in the positive x direction with speed c. Define $\xi \equiv x - ct$. Substituting $f(\xi)$ into Burghers equation yields $f_{\xi\xi} - f f_\xi = -c f_\xi$. This ordinary differential equation can be readily integrated. After imposing the boundary conditions, one gets $u(x, t) = \frac{u_0}{2}[1 - \tanh(u_0(x - ct)/4 + A)]$, where A is an arbitrary constant. Similarly, the traveling wave solutions of the KdV and the modified KdV equations are $u(x, t) = -\frac{c}{2}\text{sech}^2[\sqrt{c}(x - ct)/2 + A]$ and $u(x, t) = \pm\sqrt{c}\,\text{sech}[\sqrt{c}(x - ct) + A]$ respectively.

Chapter 7.

Problem 7.1: The superpotential $W(x) = A\sin x$ has period 2π. The corresponding partner potentials are $V_1(x) = \sin^2 x - \cos x$ and $V_2(x) = \sin^2 x + \cos x$. Since both $\exp(\pm\int W dx) = \exp(\mp A\cos x)$ are normalizable, and eq. (7.4) is satisfied, this is an example of unbroken SUSY. Note that $V_2(x) = V_1(x + \pi)$, which means that the partner potentials are self-isospectral.

Problem 7.2: The Lamé potential $V_1(x) = 2m\,\text{sn}^2(x, m) - m$ has period $2K(m)$. The function $\psi_0(x) = \text{dn}(x, m)$ also has period $2K(m)$ and no nodes. It is a zero energy band edge, since it is easy to check that $-\frac{d^2}{dx^2}\psi_0(x) + V_1(x)\psi_0(x) = 0$. The superpotential is $W(x) = -\psi_0'/\psi_0 = m\,\text{sn}(x, m)\,\text{cn}(x, m)/\text{dn}(x, m)$ and the SUSY partner potentials $V_{1,2}(x) =$

$W^2 \mp W'$ are given by

$$V_1(x) = 2m \ \text{sn}^2(x, m) - m \ ,$$

$$V_2(x) = \frac{2m^2 \ \text{sn}^2(x, m) \ \text{cn}^2(x, m)}{\text{dn}^2(x, m)} - m \ \text{sn}^2(x, m) + m \ \text{cn}^2(x, m) \ .$$

Using the relation $\text{sn}(x+K(m), m) = \text{cn}(x, m)/\text{dn}(x, m)$ and algebraic identities involving Jacobi elliptic functions, one can show that $V_1(x+K(m)) = V_2(x)$, thus establishing the self-isospectrality property.

Problem 7.3: The associated Lamé potential under consideration is

$$V_1(x) = 2m \ \text{sn}^2 x + 2m \ \text{cn}^2 x/\text{dn}^2 x \ .$$

It has period $K(m)$. Choosing $a = 1$, $n = 3$ in Table 7.3 immediately yields the desired three eigenvalues and corresponding eigenfunctions. The lowest energy is $E_0 = 2 + m - 2\sqrt{1 - m}$ with a nodeless wave function $\psi_0 = \text{dn}(x, m) + \sqrt{1 - m}/\text{dn}(x, m)$ of period $K(m)$. The next two energy levels are at $E_1 = 2 + m + 2\sqrt{1 - m}$ and $E_2 = 4$ with wave functions $\psi_1 = \text{dn}(x, m) - \sqrt{1 - m}/\text{dn}(x, m)$ and $\psi_2 = \text{sn}(x, m) \ \text{cn}(x, m)/\text{dn}(x, m)$ These wave functions have a period $2K(m)$ and one node in every period.

Problem 7.4: The associated Lamé potential for $(a, b) = (2, 1)$ is $6m \ \text{sn}^2 x$ $+2m \ \text{cn}^2 x/\text{dn}^2 x$. It has period $2K(m)$. Choosing $a = 2$, $n = 1$ in Table 7.3 gives the ground state energy $E_0 = 4m$ and wave function $\psi_0 = \text{dn}^2(x, m)$ As expected, ψ_0 is nodeless and has period $2K(m)$. The corresponding superpotential is

$$W(x) = -\psi_0'/\psi_0 = 2m \ \text{sn}(x, m) \ \text{cn}(x, m)/\text{dn}(x, m) \ .$$

The SUSY partner potentials are

$$V_1(x) = 6m \ \text{sn}^2 x + 2m \ \text{cn}^2 x/\text{dn}^2 x - 4m \ ,$$

$$V_2(x) = 2m \ \text{sn}^2 x + 6m \ \text{cn}^2 x/\text{dn}^2 x - 4m \ .$$

Clearly, $V_2(x)$ is just $V_1(x)$ shifted by a half-period. Hence the two potentials are self-isospectral.

Chapter 8.

Problem 8.1: The WKB integral $\int_{x_L}^{x_R} dx \sqrt{E - V(x)}$ for the harmonic oscillator potential $V(x) = \omega^2 x^2/4$ can be easily evaluated to be $\pi E/\omega$, which gives the energy levels $E_n^{WKB} = (n + \frac{1}{2})\omega$, in agreement with the exact answer. For the Morse potential $V(x) = A^2 + B^2 e^{-2\alpha x} - 2B(A + \alpha/2)e^{-\alpha x}$, the WKB integral is $\frac{\pi}{\alpha}[(A + \alpha/2) - \sqrt{A^2 - E}]$, leading to the energy levels $E_n^{WKB} = A^2 - (A - n\alpha)^2$, which is again the exact answer. Similar calculations using $W^2(x)$ and the SWKB approximation also give exact energy levels for the harmonic oscillator and Morse potentials.

Problem 8.2: The potential $V(x) = A^2 - A(A+1)\text{sech}^2 x$ is a special case $(B = 0)$ of the Rosen-Morse II potential. This potential is shape invariant, and hence analytically solvable. The exact eigenenergies are $E_n = A^2 - (A - n)^2$. The WKB approximation is

$$2 \int_0^{x_R} dx \sqrt{[E_n^{WKB} - A^2 + A(A + 1)\text{sech}^2 x]} = (n + \frac{1}{2})\pi \ ,$$

where the classical turning point x_R is determined from $V(x_R) = E_n^{WKB}$. The integral can be evaluated after a change of variables $y = \tanh x$. Algebraic simplification then yields the result

$$E_n^{WKB} = 2(n + \frac{1}{2})\sqrt{A(A + 1)} - (n + \frac{1}{2})^2 \ .$$

For very large values of n and A, the WKB approximation clearly reduces to the exact eigenenergies. For $A = 4$, the WKB energies are $4.22, 11.17, 16.11, 19.05$. These are to be compared with the exact values $0, 7, 12, 15$, and the accuracy is clearly very poor.

The superpotential which corresponds to $V(x)$ is $W = A \tanh x$. The SWKB approximation is

$$2 \int_0^{\tilde{x}_R} dx \sqrt{[E_n^{SWKB} - A^2 \tanh^2 x]} = n\pi \ ,$$

where the classical turning points satisfy $V(\tilde{x}_R) = E_n^{SWKB}$. After considerable algebraic simplification, one gets $E_n^{SWKB} = 2An - n^2$, which agrees with the exact result!

Problem 8.3: The energy levels of the Coulomb potential in the WKB approximation are given by

$$\int_{r_L}^{r_R} dr \sqrt{[E_n^{WKB} + e^2/r - l(l+1)/r^2]} = (n + \frac{1}{2})\pi ,$$

where the classical turning points r_L, r_R are the two solutions of the quadratic equation $E_n^{WKB} + e^2/r - l(l+1)/r^2 = 0$. Upon simplification, this yields

$$E_n^{WKB} = \frac{-e^4}{4[n + \sqrt{l(l+1)} + \frac{1}{2}]^2} .$$

The Langer correction replaces $\sqrt{l(l+1)}$ by $l + \frac{1}{2}$, thereby yielding exact energy levels.

The superpotential $W(r) = \frac{e^2}{2(l+1)} - \frac{(l+1)}{r}$ gives the potential $\tilde{V}(r) = \frac{-e^2}{r} + \frac{l(l+1)}{r^2} + \frac{e^4}{4l(l+1)}$. The SWKB condition gives the energy levels of $\tilde{V}(r)$ to be

$$E_n^{SWKB} = \frac{e^4}{4l(l+1)} - \frac{e^4}{4(n+l+1)^2} ,$$

which corresponds to the exact spectrum for the three dimensional Coulomb potential.

Problem 8.4: The WKB energy levels are given by

$$2 \int_0^{[E_n/\alpha]^{1/q}} \sqrt{E_n - \alpha x^q} \, dx = (n + \frac{1}{2})\pi .$$

Making a change of variables $y = \sqrt{E_n - \alpha x^q}$, we obtain

$$\frac{4}{q\alpha^{1/q}} \int_0^{\sqrt{E_n}} dy y^2 (E_n - y^2)^{(1-q)/q} = (n + \frac{1}{2})\pi .$$

The energy dependence can be scaled out by a further change of variables $u = y^2/E_n$. The remaining integral is a beta function $B(\frac{3}{2}, q^{-1})$. After some algebraic simplification, one gets the desired result for the WKB energy levels

$$E_n = \left\{ \frac{(n+\frac{1}{2})\hbar\alpha^{1/q}\Gamma(\frac{3}{2}+q^{-1})}{4(2m)^{1/2}\Gamma(\frac{3}{2})\Gamma(1+q^{-1})} \right\}^{2q/(q+2)} .$$

Chapter 9.

Problem 9.1: a) The exact ground state wave function is

$$\psi_0 = N\exp[-x^4/4] ,$$

and the ground state energy is $E_0 = 0$.
b) The ground state energy functional is

$$E_0(\rho,n) \times \Gamma(\frac{1}{2n}) = \frac{n^2}{2\rho}\Gamma(\frac{4n-1}{2n}) + \frac{\rho^3}{2}\Gamma(\frac{7}{2n}) \pm \frac{3\rho}{2}\Gamma(\frac{3}{2n}) .$$

c) For H_1 we have for $n = 2, \rho = \sqrt{2}$, and for $n = 1, \rho = \left(\frac{2}{15}(1+\sqrt{6})\right)^{1/2}$.
For H_2 we have for $n = 2, \rho = .816497$, and for $n = 1, \rho = \left(\frac{2}{15}(-1+\sqrt{6})\right)^{1/2}$.
d) For H_1 we have for $n = 2, E = 0$ and for $n = 1, E = .152$. For H_2 we
have for $n = 2, E = 1.104$ and for $n = 1, E = .978$.

Problem 9.2: For this problem with $2V(x) = x^6 - \alpha x^2$ with the choices
$\alpha = 7, 11$ (which is a double well potential) the exact wave function is of
the form $e^{-x^4/4}$ multiplied by a polynomial which is quadratic and quartic
respectively for the choices $\alpha = 7, 11$. Since the exact wave function has
two regimes of large support not at the origin, we do not expect our form
of the trial wave function to be that accurate. However we will find that
taking a non-Gaussian trial wave function can definitely improve the value
of the ground state energy estimate over a Gaussian. The exact answers
are $E_0 = -\sqrt{2}$ for $\alpha = 7$ and $E_0 = -4$ for $\alpha = 11$. The variational result is

$$E_0(\rho,n) \times \Gamma(\frac{1}{2n}) = \frac{n^2}{2\rho}\Gamma(\frac{4n-1}{2n}) + \frac{\rho^3}{2}\Gamma(\frac{7}{2n}) - \frac{\alpha\rho}{2}\Gamma(\frac{3}{2n}) .$$

Setting $n = 1, 2$ and minimizing this functional we obtain for the choice $\alpha = 7$ the following: for $n = 1 : \rho = 0.8614, E_0 = -0.618$; for $n = 2 : \rho = 1.86892, E_0 = -1.11221$. These need to be compared to $E_0 = -1.414$. Instead for the value $\alpha = 11$, we find for $n = 1 : \rho = 1.0303, E_0 = -1.56535$; and for $n = 2 : \rho = 2.2689, E_0 = -2.5139$. These need to be compared to $E_0 = -4.0$.

Problem 9.3: This problem is self explanatory.

Problem 9.4: The radial Schrödinger equation in N dimensions is

$$\left[-\frac{\hbar^2}{2m}\frac{d^2}{dr^2} + Ar^\nu + \frac{(N + 2l - 1)(N + 2l - 3)\hbar^2}{8mr^2} \right] \phi = E\phi .$$

The remaining steps and results are described in the problem.

Problem 9.5: Note that, as expected from our choice of $\bar{k}$, only the leading term in the result for E_n survives for $\nu = -1$ and $\nu = 2$. The computation of the $O(\bar{k}^{-2})$ terms is quite involved. The interested student can find details in U. Sukhatme and T. Imbo, Phys. Rev. **D28** (1983) 418-420.

Appendix C.

Problem C1: In logarithmic perturbation theory, the procedure is to start with the unperturbed normalized ground state wave function and energy, $\psi_0(x)$ and E_0, and then successively compute the quantities $E_1, C_1, E_2, C_2, E_3, C_3, \ldots$ using the formulas given in the text. For this problem, the unperturbed starting points are

$$E_0 = \frac{\hbar\omega}{2} , \quad \psi_0(x) = \left[\frac{\alpha}{\pi}\right]^{1/4} e^{-\alpha x^2/2} , \quad \alpha \equiv \frac{m\omega}{\hbar} .$$

The first-order energy correction E_1 vanishes, since one has an antisymmetric integrand and symmetric limits of integration. The quantity $C_1(x)$ is given by

$$C_1(x) = \frac{2me^{\alpha x^2}}{\hbar^2} \int_{-\infty}^{x} (-e\mathcal{E}x) e^{-\alpha x^2} dx = \frac{me\mathcal{E}}{\hbar^2 \alpha} .$$

The second-order energy correction E_2 is

$$E_2 = -\frac{\hbar^2}{2m} \int_{-\infty}^{\infty} \left[\frac{me\mathcal{E}}{\hbar^2 \alpha}\right]^2 \left[\frac{\alpha}{\pi}\right]^{1/2} e^{-\alpha x^2} \, dx = -\frac{e^2 \mathcal{E}^2}{2m\omega^2} .$$

The quantity $C_2(x)$ is given by

$$C_2(x) = -e^{\alpha x^2} \int_{-\infty}^{x} \left[\frac{2m}{\hbar^2} E_2 - C_1^2(x)\right] e^{-\alpha x^2} \, dx = 0 .$$

Since C_2 vanishes, so does E_3 as well as all higher-order corrections. The final well-known result is

$$E = \frac{\hbar\omega}{2} - \frac{e^2 \mathcal{E}^2}{2m\omega^2} ,$$

which is the exact answer.

Problem C2: The unperturbed normalized ground state wave function and energy are

$$E_0 = -\frac{e^2}{2a_0} , \quad u_0(r) = \left[\frac{4}{a_0^3}\right]^{1/2} r e^{-r/a_0} , \quad \int_0^{\infty} |u_0(r)|^2 \, dr = 1 .$$

Using the logarithmic perturbation theory formulas, performing the necessary integrals, and discarding terms of $O(\eta^4)$, one gets

$$E_1 = E_0 \left[-\frac{\eta^2}{3} + \frac{\eta^3}{6} + O(\eta^4)\right] , \quad E_2 = E_0 \left[\frac{2\eta^3}{45} + O(\eta^4)\right] ,$$

$$E_{n>2} = E_0 \, O(\eta^4) .$$

The final result for the ground state energy is

$$E = E_0[1 + \beta_2 \eta^2 + \beta_3 \eta^3 + O(\eta^4)] ,$$

with $\beta_2 = -\frac{1}{3}$ and $\beta_3 = \frac{19}{90}$.

Index